Decoding Persistent Depression

Book One - Mysteries and Mindsets

Dr. Roger Di Pietro, Psy.D.
with special contribution from
Dr. Harold H. Mosak, Ph.D., ABPP

© 2018 Dr. Roger Di Pietro, Psy.D.
All rights reserved.
ISBN: 978-1-387-80068-1

Other books by Dr. Roger Di Pietro, Psy.D.

Early Recollections: Interpretative Method and Application
(Co-authored with Dr. Harold H. Mosak, Ph.D., ABPP)

The Depression Code: Deciphering the Purposes of Neurotic Depression

The Anxiety Code: Deciphering the Purposes of Neurotic Anxiety

Dedication

Dr. Harold H. Mosak, Ph.D., ABPP
1921-2018
Friend, mentor, author, teacher extraordinaire
who encouraged this project.

Acknowledgments

Thank you to my family.

Dr. Joseph Fullone, M.D. for his comments and assistance.

My sincerest appreciation goes to those inspiring individuals
who through courageous and determined effort
conquer their symptoms.

Introduction

Depressive disorders are the most commonly reported psychological symptom constellations. In fact, it may be helpful for you to view depression as a large diagnostic family, with several siblings that tend to resemble each other.

This book investigates one specific depressive diagnosis: Dysthymia, which is also known by its many aliases: persistent depressive disorder, dysthymic disorder, and neurotic depression. (Terms I'll use interchangeably to avoid the risk of grating repetition.)

Briefly, it's a long-term condition (two years or more) and less severe than major depression.

Before you get started, you may wish to know about my academic instruction and clinical point of view, as they – along with theory and research from various perspectives and disciplines – shape and support the pages that follow.

My undergraduate training consisted of a triple major, with psychology my primary degree. In addition, I have two master's degrees in psychology.

The first is a Master of Arts in the Social Sciences (Behavior Modification). Training in behavior therapy deals primarily with the individual. It does not delve into psychoanalysis or personality, nor does it deeply examine the social component of an individual's functioning. While I continue to embrace the strengths of that theoretical and treatment orientation, after graduation I craved for a wider, more encyclopedic view of human behavior.

I decided to study in Europe.

I earned a Master of Science in Social Psychology from the London School of Economics and Political Science, University of London. This experience broadened my comprehension and appreciation of subjective perception and the social influences on an individual. My education came from class work as well as my interactions with students and professors from different nations and diverse cultural backgrounds. The importance of being able to see through many lenses to get a panoramic perspective was a point well made.

After London, I moved to Chicago and graduated from the Adler School of Professional Psychology (known as Adler University after January 2015) with a doctorate in clinical psychology and having earned membership in Psi Chi, the International Honor Society in Psychology.

Although this book encompasses many theoretical orientations and clinical practices, it's primarily based on Adlerian theory, research, and treatment. With that in mind, it's useful for you to have a glimpse of Dr. Adler's work as well as that from some prominent Adlerian practitioners.

Dr. Alfred Adler was one of the three founding fathers of psychotherapy and counseling along with Dr. Carl Jung and Dr. Sigmund Freud. Adler broke away from his colleague Freud, left the psychoanalytic model of treatment, and created his own school of psychotherapy: *Individual Psychology*. He proposed that people are unique, active, goal oriented, creative, and social.

Adler championed the idea that people are social beings, influenced by interaction and shared desires (e.g., wanting to be competent, loved, accepted, valued, and in harmony with others) rather than – like Freud believed – driven by biological urges and instincts.[1]

(Does this age-old argument sound a bit familiar? One current variant of this discussion is between a chemical imbalance [or other innate issues, such as genetics, or being hardwired for symptoms], and the role of individual choice and social factors in developing depression.)

Adler influenced many psychological assessment and treatment orientations (e.g., cognitive-behavioral therapy), directly and indirectly, during the last century and to this day.[2] Although being an early contributor to the field, his work is in agreement with the current biopsychosocial approach in comprehending behavior and treating psychopathology.[3]

In addition, he emphasized equality and democracy, and avoided the traditional psychoanalytic couch, preferring clinicians and patients sit in chairs and work collaboratively – a practice that remains today's standard.

Adler lectured in Europe and the United States. He insisted that his ideas did not have to live solely in academic and clinical settings, believing that laypersons could find his concepts and practices useful.

Accordingly, he spoke to and wrote for the public, enabling a widespread distribution of information. Adler shared his understanding of psychological symptoms to aid and advance assessment and prevention.[4]

Under Adler's firsthand tutelage was Dr. Rudolf Dreikurs, a notable psychiatrist in his own right, who – among other impressive contributions – expanded and applied Individual Psychology to parent training as well as assessing and treating children.

In 1952, Dreikurs and others founded what is now Adler University in Chicago. One of these co-founders was his student and eventual colleague, co-therapist, and co-author, Dr. Harold H. Mosak.

Dr. Mosak is a psychologist who lectured at more than 70 schools, colleges, and universities around the globe, published a combination of more than 130 articles and books, and was in private practice for nearly six decades. His contribution to the field of Adlerian psychology is virtually unparalleled. He was my supervisor in graduate school, encouraged this book, and discovered the personality factors I've nicknamed the Depression Code.

However, this isn't our first collaboration.

During my internship, he generously asked me to co-author a textbook on the history and use of childhood memories as a personality assessment: *Early Recollections: Interpretative Method and Application.*

After earning my doctorate in clinical psychology in Chicago, I completed my postdoctoral training in Marriage and Family Therapy at the University of Rochester Strong Medical Center. Currently, I'm a clinical psychologist in private practice in Upstate New York.

Comparable to Adler's presentation style, this book is written as a friendly, instructive, investigative dialogue, full of questions and easy-to-understand examples, thus similar to therapy or a classroom interaction. Familiar, manageable terms are used instead of clinical, archaic, or uncommon expressions, and endnotes discreetly point to supporting books and articles.

This text explores individual choice, social functioning, and psychological underpinnings in the development and maintenance of dysthymia. The goal is to palatably provide a point of view that may inspire dialogue, dispel myths, increase comprehension, improve care, and generate research. Given the vast amount of information, it may be best to consume it as slowly and deliberately as you would rich chocolate.

This is the first book in a series which is based on and expands a previous text: *The Depression Code: Deciphering the Purposes of Neurotic Depression.*

While the presented information has been used to successfully treat patients, it isn't to be used for an official diagnosis. Nor, for reasons you'll discover, is this a step-by-step self-help manual. Yet, the material within can assist in developing understanding – for symptomatic persons, educators, as well as healthcare providers – and employed to reduce symptoms.

Clues along the way will help you eliminate distracting, contradictory, misleading, and countertherapeutic concepts. As you gather illuminating evidence, you'll be better able to comprehend and resolve an intriguing – and seemingly impenetrable – mystery, with all the usual suspects (and perhaps a few surprises): How can persistent depression arise and why might it persist?

- Dr. Roger Di Pietro, Psy.D.

Table of Contents

DEDICATION 4
ACKNOWLEDGMENTS... 5
INTRODUCTION 6
CAUSE AND EFFECT 12
MINDSETS AND TREATMENTS 16
RISKS, DANGERS, AND DOUBTS 20
MEMORIES 21
INSIGHT 32
COMPLEXITY 34
GENETICS AND BEING "HARDWIRED" 36
CHEMICAL IMBALANCE 41
CHEMICAL IMBALANCE QUESTIONS 44
SEARCHING FOR A CAUSE 50
SYMPTOM FUNCTIONS 53
SYMPTOM SUPPRESSION 60
ANTIBIOTICS, PAINKILLERS, AND E-MAIL FILTERS 61
TELLING ANALOGIES 64
POINTS TO CONSIDER 69
MODIFYING MINDSETS 71
CLINICAL CLUES 75
GROWTH MODEL 79
PERCEIVING PEOPLE AS PASSIVE 82
GOAL ORIENTATION 84
LIFE TASKS 91
INFERIORITY 95
COMPETENCE 101
LIFELONG SOCIALIZATION 106
SOCIAL INTEREST 109
SOCIAL INFLUENCE 112
NOTHING HURTS MORE THAN LOVE LOST 118
NOTEBOOK OF CLUES AND EVIDENCE 121
COMMON SENSE 126
COMMON SENSE MAY DEFINE MENTAL WELLNESS 133
PROBLEMATIC COMMON SENSE 134
COMMON SENSE SOLUTIONS 137
SELF-DETERMINATION 140
PSYCHIATRY AND PSYCHOLOGY: HARDWARE AND SOFTWARE 145
PERSONALITY 153
LIFE STYLE CONVICTIONS 155
LIFE STYLE CONVICTIONS CAN BE STRONG AND INFLEXIBLE 161
SUBJECTIVE CONVICTIONS 167
DIFFERENT CONVICTIONS 169
BIODIVERSITY CLUES 171

FICTIONAL FINALISM	177
"REAL" MAN AND "REAL" WOMAN	179
ADAPT, SURVIVE, AND THRIVE	181
REVEALING CHOICES	186
NOTEBOOK OF CLUES AND EVIDENCE, PART TWO	191
PRIVATE LOGIC	196
THE QUEST FOR COMMON CONVICTIONS	201
THE DEPRESSION CODE	204
THE NEED TO GET	204
THE NEED TO BE IN CONTROL	205
THE NEED TO BE GOOD, PERFECT, RIGHT	209
PROBLEMS WITH THE NEED TO GET	212
PROBLEMS WITH THE NEED TO BE IN CONTROL	217
PROBLEMS WITH THE NEED TO BE GOOD, PERFECT, RIGHT	225
INADEQUATE AND TROUBLESOME CONVICTIONS	230
INSUFFICIENT AND INCORRECT	230
COMPOUNDING	230
CONFLICTING	232
CANDY BAR CONFLICT	232
COOKIE JAR CONFLICT	233
STOPLIGHT CONFLICT	234
COMMON SENSE VS. PRIVATE LOGIC	235
CLIFFHANGER	239
REFERENCES	242

Cause and Effect

Depression: A colorless, seemingly unshakable shadow, so saturated with heavy thoughts that it clings tirelessly, blanketing its sufferers with oppressive weight, lacking warmth or comfort.

Millions of Americans per year experience depression, with up to 6% of the American population at some point in their lifetime.[5] In addition, it appears to be the leading cause of disability around the globe.[6] Long-term depression lessens people's quality of life and increases the probability of cognitive impairment for older individuals.[7]

Depression is more common and incapacitating than people tend to recognize.

Intriguingly, women report symptoms more often than men.[8]

But this doesn't mean women are inferior to men cognitively, biologically, chemically, physiologically, or genetically.

For instance, some men might choose to "tough out" their symptoms – even if they don't go away. Others want to avoid investigating painful thoughts and experiences. Some strictly forbid themselves from talking about their feelings and faults because they don't want to be seen as "weak", be deemed failures, or express their emotions in front of others.

So, the difference in reporting may be due to women's *healthier* attitude toward their well-being (e.g., prioritizing their health, going to their doctors more frequently, less worried about what others think).[9]

This example serves as an illustrative reminder that data needs to be very carefully interpreted, otherwise risk being inadvertently misleading. As you'll see, there are a number of twisting dark alleys and enticing trap doors along the path to demystifying depression.

Moving forward, you should also know that "depression" is a general term. Therefore, it may be helpful to view it almost as if it were a surname – a family of related symptoms that lives by various diagnostic names, such as major depressive disorder (unipolar depression), dysthymia, bipolar I and bipolar II disorder (manic depression), cyclothymia, depression with seasonal pattern (seasonal affective disorder), recurrent depressive disorder, and mood disorder due to another medical condition.[10]

This book focuses on one specific depressive diagnosis: dysthymia, which is also referred to as dysthymic disorder, neurotic depression, as well as persistent depressive disorder.[11] (Diagnostic names that will be used interchangeably; simply – and somewhat mercifully – to avoid repeating the same term.)

Briefly, *it's a relatively consistent mild form of depression that's often accompanied by anxiety.* Symptoms include some or all of the following: fatigue, persistent sadness, notable increases or decreases in appetite, an inability to sleep (or remain asleep) as well as oversleeping, reduced ability to concentrate, indecisiveness, and despondency. These symptoms must be present for at least two years and cause significant distress and/or impair functioning in one or more areas of life.[12]

Dysthymic disorder is a unique diagnosis, and can co-exist with major depressive disorder, in a condition known as double depression.[13] (Think of this like two roommates that can aggravate and antagonize each other *and* create a particularly difficult time for their live-in landlord.)

Depression can ruthlessly and relentlessly sap people's strength, foil attempts at success, corrupt mood, poison concentration, arrest movement toward stated goals, obstruct opportunity, endanger careers, challenge friendships, and sabotage relationships.

When you think about it, how could symptoms such as anger, sadness, low self-esteem, reduced energy, unpredictable mood swings, irritability, agitation, decreased ability to experience pleasure, as well as feelings of worthlessness, guilt, hopelessness, and helplessness *not* be troublesome?

There are a number of other factors that make matters worse; for example, an inexplicable timing of symptoms that erupt without apparent provocation or warning. Moreover, symptom duration may vary to the point of bewilderment, with neither depressed individuals, nor those around them, able to accurately forecast symptoms persistence or intensity.

For these reasons, neurotic depression can appear unnervingly vague, illogical, unpredictable, impulsive, uncontrollable (e.g., beyond medication or therapy), overwhelming, and prone to hinder or halt people's ambitions.

Often, people think of the following reasons why depression exists.
- Genetics
- Brain injury
- Malnutrition
- Low blood sugar
- Sleep deprivation
- Improper parenting
- A loved one's death
- Chemical imbalance
- Recent unemployment
- Psychological or physical trauma
- "Bad wiring" (problematic neurological structures)
- Cloudy weather and a lack of full spectrum lighting
- The unanticipated and undesired end of a relationship

Notice that each of the above reasons for depression share an underlying mindset: *cause and effect*. In other words, a situation, experience, or condition causes depression. This cause and effect mindset is common in understanding. Just consider how pushing an object makes it fall, a physical impact results in a bruise, heavy rains lead to a flooded basement, or how bacteria and viruses produce an infection. People are accustomed to think in causal terms. After all, life provides countless examples, medical and otherwise, that serve as proof of its legitimacy. The cause and effect point of view has become so entrenched and popular that it may render itself invisible and accepted as an inescapable part of life.

So, it seems rather natural and logical to assume that some condition or event produces depression. In fact, various conceptions of depression assert a genetic, biological, chemical, social, or physiological origin. And people often – and for some individuals, exclusively – perceive depression as caused by something.

But there can be a downside.

Often accepted as a well-established certainty, the cause and effect mindset can be a stealthily-influential starting point for assessment and treatment that risks going undetected, and therefore, unexamined. Therefore, to more fully comprehend depression, you must acknowledge its theoretical underpinnings and how the cause and effect perspective has guided mental health assessment and treatment in many ways.

Mindsets and Treatments

Attempts at understanding and treating psychological symptoms have a long – and somewhat curious – history.

To start, consider how children too young for abstract thought may explain why something complex occurred with the naive (and incorrect) answer, "Magic".

Adults tend to do this as well.

When they lack a logical or scientific answer, or have difficulty grasping complicated, intricately-interwoven factors, they may place explanation in the simpler, concealed realm of the supernatural.

People can effortlessly leapfrog to an uncomplicated conclusion without getting near the messy, taxing cognition in-between. The embryonic stages of psychological assessment and treatment were not immune to this phenomenon.

For example, ancient and outdated beliefs held that demonic infestation caused mental distress. As a result, people with unhealthy psychological conditions and behavioral peculiarities were viewed as innocent victims that fell prey to unseen forces or demons.[14]

This conception may appear ludicrous to the educated eyes of modern science. However, there's a thinly-veiled, but boldly-etched clue to these mystical origins. The literal translation of the word "psychology" is "study of the soul".[15]

Given a metaphysical mindset, witch doctors and similar others misguidedly attempted to evict the (assumed) resident demons through rituals, such as sacrifices, exorcisms, chants, and drilling holes in the heads of those afflicted in a process known as trepanation.[16] These painful and primitive practices are best left dormant in the safe distance of history.

Later, immorality was seen as the transgression that got the attention of the Antichrist, the Devil, fallen angels, werewolves, tarantulas, and dybbuks that wickedly inflicted psychological suffering.[17]

And it's important to know how that perspective influenced assessment and treatment.

Sadly, yet somewhat predictably, once people were perceived as morally responsible for their mental condition, punishment (!) was added to the list of existing treatments.[18] This, as you can easily imagine, could be immensely countertherapeutic.

In the Western world, supernatural perspectives of psychological symptoms have been mostly cast out, and the barren belief that immorality causes psychological symptoms has been largely abandoned.

Another perspective was cultivated in their place.

More recently, psychological symptoms came to be seen as disease-based mental illness and were brought into the medical domain, where physicians applied their conceptual framework for assessing, comprehending, and treating psychiatric and psychological symptoms.[19]

The disease model transplanted many features from the medical field to the psychological one, such as diagnosis, viewing the person (labeled "patient") as sick, extensive testing, focusing on and attempting to eliminate symptoms, searching for causes and cures, and the use of asylums and medications. (The practice of using pre-existing mindsets and borrowed terminology is rather common. For instance, modern vehicles are measured in *horse*power.)

This perspective perceives symptoms as indicators of something broken or infected, and therefore in need of fixing. This is similar to how pain indicates a broken bone, which then needs to be set. It's a model that not only seeks causes, but also means by which to alleviate their effects.[20]

Given the medical model's ability to effectively and humanely remedy so many physical illnesses, it shouldn't be a surprise that it's commonly employed in the assessment and treatment of psychological symptoms. Indeed, those who view depression through a medical lens tend to seek an assortment of genetic, biological, neuronal, physiological, or chemical therapies. Perhaps the most common are the use of psychotropic medications, such as antidepressants.

Pharmacological treatments influence brain activity, for example, a problem in neurotransmitter production and functioning, or with neuronal action.[21] Indeed, symptomatic persons with relatively little effort – taking a pill – can quickly influence their mood. Yet, consider when using antidepressants and experiencing swift symptom reduction reinforce a medical mindset of symptom origin *even when it's inapplicable.*

Of the many offshoots of the well-rooted cause and effect tree, the medical model is often seen as the strongest and most fruitful branch, with an ability to decrease or eliminate symptoms by whittling the comprehension and treatment of depression down to mechanistic causes and effects (e.g., neurotransmitter levels, neurological functioning, physiological actions).

Also, non-medical conditions are seen as triggering depression.

When psychological trauma or unhealthy ways of thinking are perceived as causing depression, people may enter counseling or psychotherapy (e.g., supportive, cognitive-behavioral therapy). These forms of treatment focus on individual perspective and choice, attempt to improve people's intrapersonal and interpersonal functioning, and change thinking and behavior patterns. These approaches can be quite effective at reducing symptoms, as well as prevent their recurrence.

But there are some inherent challenges.

First, just like people cannot learn to play the piano or adopt a new language in just a few classes, the process of acquiring insight, taking a different viewpoint, and instituting new rules of behavior can be a lengthy and demanding one. Some individuals prefer the convenience and time-saving practice of taking medication.

Second, how and what people think construct their reality. Therefore, they may be especially reluctant to change their perspective – *even when it's related to agonizing symptoms* – as that requires a possibly unsettling alteration of their sense of reality.

Third, symptoms can painfully prohibit doing what's healthy (e.g., socialize, work, be in a relationship, self-care, pay attention in therapy). This makes recovery that much more difficult.

With that in mind, think about those who take analgesics (aka painkillers) to do physical therapy, which can be a painful, but ultimately healthy, activity to better understand a commonly-employed solution to these challenges.

Antidepressants are frequently used in conjunction with psychotherapy to assist symptom reduction or alleviation while psychotherapy addresses the events, conditions, thoughts, and behavior patterns that are believed to generate symptoms.[22]

Like the teamwork seen with anesthesia and surgery, psychiatry and psychology can be employed cooperatively to achieve treatment goals.

As the preceding isn't an exhaustive list of mindsets and interventions, think about how else people may conceptualize and treat depression.

To start, consider those whose faith lies in alternative methods, or who have failed to be saved by mainstream therapies and subsequently renounce them.

They often seek eclectic or unorthodox treatments such as yoga, light therapy, aromatherapy, dietary supplements, meditation, finding distractions in hobbies, associating with encouraging individuals, massage, exercise, acupuncture, or seeking solace and acceptance in religion or spirituality.

Others may unsoundly try to self-medicate with alcohol, tobacco, caffeine, "comfort food", or illegal drugs.

Some of these activities may intensify symptoms, while others can provide mild relief or temporary symptom reduction, but are generally ineffective for stubborn, long-term depression.

There's a crucial relationship between mindset and treatment that all of the preceding illustrate.

The ways in which people perceive psychological conditions governs how they act toward them.

As each of the aforementioned perspectives has a cause and effect framework, it's easier to understand how the causal model of depression is prevalent. For some people, however, the cause and effect mindset is indisputably the *only* approach to assessing and treating dysthymia.

Risks, Dangers, and Doubts

If people retained the belief that demonic infestation caused psychological symptoms it might be logical – and therefore a continued practice – to drill holes in individuals' heads in an effort to dispossess demons.

While you might believe that such practices are in the ancient past, despite contraindicating medical evidence, there remains a tiny fraction of the world's population who willingly undergo this process in an attempt to improve their mental health and capacity.[23]

This illustrates and reaffirms your understanding that the way in which people conceive of psychological symptoms guides their assessment and treatment actions – even to an unhealthy outcome.

This brings you to an inescapable conclusion.

Rather than uncritically and wholeheartedly accept any theory (or the assessment and treatment that grow from it), it's better to remain vigilant for any shortcomings and thorny issues.

So, it may be helpful to investigate the possible drawbacks of *only* holding a cause and effect mindset of persistent depressive disorder. Perhaps upon closer examination there could be limitations, inconsistencies, and unsatisfied questions that repeatedly pop up like tenacious weeds in a garden.

With that in mind, consider some common conceptions of what causes depression as well as their related assessment and treatment.

Memories

It's crucial for people to recall things such as, how to get home, that the stove is on, or that their newborn baby is in the backseat of the car.

In addition, consider how individuals may try to remember, for example, why they dislike a person ("Oh, yes, he insulted his wife at the office party"), why the basement door is open ("I forgot to close it after doing the laundry" [A far less frightening conclusion than believing an intruder is hiding downstairs]), or why their coat smells like smoke ("That's right, we had lunch at the pub").

These realities lead to particular conclusions.

People can productively search their memories to recall important facts as well as find out why something occurred or exists. Consequently, it's a process that can be inadvertently, intermittently, and regularly reinforced throughout life and in countless situations as a means of understanding and predicting.

Consider how this practice and understanding relate to depression.

Contemplate how some individuals cite memories of one or more childhood incidents they believe caused their symptom, for example:

"I remember this one hot summer's day when all of the other neighborhood kids said that they were tired or hungry and were going home. As there would be no one left to play with, I went home. But, as I entered my house, I saw all of them go to my neighbor's home. *He had a swimming pool*. They knew I didn't swim. They ditched me so they could all be in the pool together without me. *Others lie to me and abandon me*. That was the day it started, and it's been that way my entire life! I've been sad ever since."

Now, think about how a memory-related cause and effect mindset of symptoms may influence which therapeutic paths people take.

They may select treatment that employs various techniques to soothe sad or upsetting memories (e.g., "It's okay, that was long ago.", "Everyone has experienced troubling times. It's normal.", "Don't worry about the past, things will turn out fine"). Or, they may attempt to offset depressive memories by finding and focusing on positive, uplifting recollections. Such practices can be as logical, familiar, and useful as hanging out with good friends after a bad day.

Combing through memories to recollect or determine something can be a convenient, beneficial, and life-long habit that starts in childhood. However, risks arise when people search for memories to determine why their depressive symptoms exist as well as for guidance on how to treat them.

As hunting for and dissecting memories for explanation and direction can be regularly practiced and intermittently reinforced, it can – like many other well-learned habits – be used routinely, forcefully, and virtually instinctively. This can make it nearly invisible in its execution, therefore possibly exempt from investigation that may reveal any problematic issues and imperfections.

Also, think about what people may conclude when they sift through their memory with relentless determination until they remember something (e.g., where they left their house keys).

Some view their memory as if it were an exhaustive reference library (e.g., a "mind palace") with eternally-enduring, archived recollections able to be read after a dedicated search.

Consider how this belief may bias depressed individuals' actions.

Given a strong, repeatedly proven, and life-long reliance on their memory, combined with their conviction that something occurred to cause their symptoms, as well as their belief that their memory is endlessly encyclopedic, they may believe that being unable to find a cause is more likely to due to forgetfulness or insufficient searching, for example, *than evidence that their belief in a cause could be mistaken*.

Some dysthymic individuals confidently assert that there *was* a critical historical event responsible for their symptoms *but* are distressingly unable to discover it – as if their mind's library had a book that was merely misplaced and required a thorough search to find it. For instance, "I'm *certain* that something happened when I was a kid to make me so depressed. But for the life of me, I don't know what it is. I *must* be repressing it!"

Knowing that, consider what therapeutic practices they may seek when they believe that unpleasant memories exist, but are so deeply buried that they're unable to resurrect and articulate them.

Tellingly, rather than surrender their belief, they may double their efforts on finding one or more recollections they believe will pinpoint the cause of their symptoms as well as suggest a means of resolving them.

Accordingly, they may attempt to brush away the occluding sands of time and attempt to disinter some elusive negative or traumatic episode that once exhumed would strip away depression's inky shroud.

For example, they may scour through childhood photos, recall old stories, or ask family and friends about the past, to find what they seek. Some enter therapy and dig through recent and remote memories to unearth a cause of their depression. They may then employ various tactics to therapeutically address those recollections.

All of this seems fair, accurate, and useful, but it may not be.

Being able to productively remember things, especially after a dedicated search, can lead to several misperceptions.

To start, the human brain is generous, but ultimately limited in how many memories it can store – impressively, it never fills up, but rather may repurpose neurons to accommodate new memories.[24] So, no matter how much effort is exerted to recover them, some experiences are irretrievable.

Some people believe that human memory is infinite and it only requires the proper technique to retrieve memories. This is false. *Memories can fade into extinction, never to be reanimated.*[25]

Next, while some memories of past events are accurate, *in general human memory can be unreliable.*[26]

What people remember may not be the photorealistic recollection of reality they believe it to be. Indeed, such trusted cognitive snapshots crack, fade, and discolor over time. Some memories fall through the cracks, evaporate without a trace, become blurred, and blend into other recollections.

For example, have you ever…

- Confused directions of how to get to a place you've been to before?
- Met a number of people at an event, but quickly forgot their names?

If so, then you're somewhat familiar with how memory can be faulty.

Now, think about the last movie you saw. Can you recollect and place each scene in order or remember all of the characters' names? Probably not (and that's just two hours in which you were recently and deliberately attentive).

Or, recall some important event from your youth at which pictures or videos were taken (e.g., birthday, graduation). Have someone else find those photos or videos and then ask you the following:
- Who was present or absent?
- What clothes were people wearing?
- Where were they sitting or standing?

While the recorded data is unaltered proof of the event that can be used to assess the accuracy of your memory, you probably already know the answer.

If you desire more evidence, consider the following research findings.

When people are shown altered photos, say from their childhood or in the news, this can lead to the creation of false memories.[27] Stated differently, a significant number of people will believe that the false photographs are accurate representations of reality. Rather than notice that the photographs are inaccurate and reject them as such, individuals may *manufacture* memories in accord with those doctored photographs.

Given the frequency and strength of people's tendency to employ memory in some advantageous way, they can have too much faith in it.

Perhaps an extreme example may illustrate more.

Think about how some depressed individuals may be so motivated to find an event or condition responsible for their symptoms, that when they cannot find a cause, they don't surrender their quest, but rather ratchet it up and attempt past life regression through hypnosis to find what they seek. However, there's no compelling evidence that past life regression is a reliable and valid practice.[28]

So, consider how memory – a routinely-trusted ally – can be such an unreliable and deceptive double agent.

First, there may be errors or difficulties in encoding an event into memory. For instance, there are biological and physiological factors such as being sleepy, ill, or on medications that make it difficult to correctly encode the experience into long-term memory. Therefore, as the "garbage in, garbage out" rule suggests, retrieval of any improperly encoded memories must be inaccurate. In other words, if you write down the wrong phone number, you'll never get in touch with the right person.

Second, people can encode memories only from their physical and psychological perspective at the time of the event. Therefore, in regard to childhood memories, for example, conditions such as children's undeveloped vocabulary, meager ability to comprehend body language, tone of voice, and behavior, as well as simple misperception based on their shorter stature and physical point of view, might limit understanding and corrupt the recording of events.

Inaccurately chronicled experiences make it impossible to retrieve authentic memories.

Third, think about how you might view all dogs if your first experience consisted of being bitten and clawed.

This one-time, emotionally-charged event might indelibly register in your mind that a specific dog – or all dogs – are lethal, therefore give you pause when contemplating approaching one.[29]

Now imagine children whose day-to-day experiences with their parents consist of watching them disagree, interact icily, make snide comments, and otherwise demonstrate mild, yet routine amounts of abrasion.

You can see how accumulating years' worth of such memories may ingrain the concept that relationships are inherently friction filled.

Basically, whether the encoding of memory is a singular dramatic incident, or a crowd of pedestrian episodes, it can generate a mood as well as fashion a lens through which people see the world. But consider how this might impact which events are encoded and how they're archived.

People may be much more likely to acknowledge those occurrences and conditions for which they're primed. (This is known as confirmation bias.[30])

For example, those who have recollections of discordant relationships may be quicker to identify marital conflict later on, which reinforces that conception, and makes it less likely to recognize instances that negate that point of view.

Simply, those actions similar to their existing template are more likely to be perceived, registered, and remembered. This may lead to particular behavior, such as avoiding marriage or apprehensively waiting for the next marital conflict (often while trying to be off the spouse's radar as much as possible).

With that in mind, consider how dysthymic individuals' perceptions of the world, others, and themselves might corrupt memory encoding.

Undeniably, when adults recall childhood memories, they do so from their current point of view.

Next, think about how happy, well-adjusted individuals can look back on embarrassing events and laugh at themselves, recollect a period in which they were devastated (e.g., after their first relationship break-up in middle school), but no longer view that as a tragedy, or reframe a difficult time as a learning experience that tested and proved their strength and determination. This leads you to the following conclusion.

People's current mindset can persuasively shape not only what memories are retrieved, but also how they present.[31]

Knowing this, what can happen with those who have an unhappy worldview?

Depressed persons can alter neutral or positive memories and make them gloomy.[32] They may darken otherwise pleasant memories, recast recollections as blue, or recall minor incidents as intolerably crushing.

Also, consider the ways low self-esteem may influence how and what people notice and record. They may be prone to seeing when they fail rather than when they succeed. Self-perception of inadequacy can occlude those positive things which are inconsistent with it. Moreover, they may use what they recall as proof that life is dismal and they're hopeless, helpless, or powerless.

Sad, sunless, depression-generating recollections can perpetuate and intensify a tainted worldview as well as lead to increasing immunity to reason and facts that may inoculate individuals from future depressive symptoms.

"Wait," you may say, "didn't you state earlier that recollecting cheerful memories can counterbalance depressing ones and alleviate symptoms?"

Reflecting upon images of happy times may lift spirits. However, *also* think about how such weightless flight can be of short duration for depressed persons. Their present mindset may override and reshape data that's contrary to their worldview. Accordingly, they may quickly observe how remembrances of their buoyant earlier selves do not mirror their current state. This may be an upsetting reminder of how far they've fallen or how cheerfulness is distant and irretrievable. Soon they lie shattered, awash in the cutting, irreparable shards of their former glory. Intriguingly (and hint, revealingly) dysthymic persons may perceive joyful recollections in a way that paradoxically worsens despair.

Fourth, recall when you disagreed with a sibling when recollecting a childhood event which each of you witnessed.

You and your sibling may have unparalleled confidence in the accuracy of the respective memory and debated furiously about which person's version is correct. This is a common occurrence.

Indeed, two people (or more) can have severely different recollections of the same episode.[33] Predictably, this is often seen in marriage and family therapy, with each person perhaps defiantly certain that what is recalled is reality, and consider the other person mistaken (or worse, delusional). As each can be stubbornly tenacious that what's remembered is accurate, there can be tremendous resistance to change, fervent argument, frustration, and caustic anger…yet, note that at least one of them is incorrect.

Fifth, there's something else you need to know about the fragility and fickleness of memories. Other people can make suggestions – or even just ask a question – that distorts or creates false memories.[34]

As an example, I've heard of stockbrokers who have greedily taken advantage of this cognitive anomaly by cold calling potential investors with phrases such as, "Do you remember when I called you last year advising you to invest in (some particular stock)?"

(At this point, it's not uncommon for the person to agree. After all, the individual has received countless phone calls and is unlikely to remember a specific one from a year ago. In addition, there are elements of social conformity – among other influential factors – that may lead to "recollecting" a purported phone call that never happened.)

The stockbroker continues, "Well, that stock has *doubled* since then. I'm calling you again today to bring your attention to another stock that I believe will do even better!"

Human memory is imprecise; therefore, it can be countertherapeutic to view recollections as a rock-solid and faultless foundation on which to build understanding or treatment.

Now, think about how other people's statements might lead to false memories as they relate to depressed individuals.

As an illustration, parents may regularly tell their dysthymic adult children of times they were unhappy in childhood and provide (accurate or inaccurate) examples (as well as *not* provide instances of when they were happy and adjusted well). You can see how this can shape and create memories and mindsets for depressed persons as well as others.

In addition, individuals may tell their dysthymic spouse that the person was depressed (or some variant of depression) at a party, sporting event, or another situation, and the spouse may believe them, or recast their memories of the event to be (perhaps more) depressing.

Sixth, people's current mindset can so strongly influence their memory that they can unknowingly fabricate recollections that reflect how they see the world, creating a false memory as if what they recall actually occurred.[35]

Individuals can unwittingly create memories of situations and events that *never happened*...but are tellingly in accord with their present mindset. This can occur without other influential factors (e.g., false photographs, mistaken others who paint dismal portraits).

Seventh, there are people who confidently state that they have an excellent memory, know that what they remember actually occurred, and that they would never, consciously or unconsciously, manufacture memories.

It's not only those with below average recall who fabricate memories.

Believe it or not, creating false recollections doesn't just occur with those of poor or average memory; *it happens equally frequently with those who have superior autobiographical memory.*[36] (Perhaps this is why eyewitness testimony can be significantly unreliable and risks being given too much credibility in court proceedings.[37])

Eighth, not all memories are corrupted or false. Individuals may accurately remember depressing events and conditions and carry them on for decades.[38] While this can validate dysthymic persons' memory of incidents, consider how searching for depressing experiences might be countertherapeutic.

Well, imagine you're on a diet. Thinking about all of the delicious, calorie-dense foods you've given up may increase your appetite for them.

And it only may be a matter of time after reflecting on and visualizing those craved items that you jailbreak your imprisoning diet and feverishly devour the food you desire.

With that in mind, consider how searching for and finding depressing memories, whether accurate or not, can shape people's mood, mindset, and actions, which can maintain, or worsen, depression.

Ninth, if individuals' current mindset influences what memories are stored and recalled can shape the presentation of what's remembered and lead to the creation of false memories, that doesn't mean childhood recollections are misleading and useless in understanding and treating dysthymic individuals.

Given that what's recalled is in accord with people's current perspective and thinking, a more appropriate use of childhood memories is *to identify how and why people's present thoughts and rules for life perpetuate their depression.*

Early recollections of childhood experiences can be used to assess people's present mindset and functioning.[39]

Those with neurotic depression may explore their memory to gain insight about the origin of their symptoms as well as a clue to their resolution. However, attempts to excavate historical causes may emphasize archeology more than psychology and can be as pointless, misguided, and perilous as trying to drive your car forward while looking in the rearview mirror.

Memories of miserable events can spark and perpetuate depression.

But there's something else you need to know about human memory and how it relates to depression. Thankfully, clues exist elsewhere in the body.

For example, odds are you've caught a cold or got a flu shot.

Note well that your body has an adaptive system of immunity in which it can remember various pathogens (e.g., disease-causing bacteria and viruses, pollen, pet hair, certain chemicals in the environment, and cancer cells) *and uses this memory to protect itself in the future*, say by launching a more specific immune attack the next time it comes in contact with a pathogen.[40]

To assist survival and thriving, people's immune system has a somewhat mathematical and mechanical memory that protectively uses past experiences to address present and future environmental threats.

This information about human memory relates to persistent depressive disorder.

Just think about how someone may say, "I'm not going to try surfing because when I was a kid we had a cottage on the lake and I nearly drowned and it took forever for people to see I was in trouble and needed help." This memory can influence action: The person avoids surfing as if the waters were teeming with hungry sharks looking to expand their diet.

Knowing that, think about why dysthymic persons might select, shape, or fabricate childhood memories.

Early recollections can be used to guide action and protect people in present and future challenges.[41]

A cause and effect mindset can prime people to use memories to comprehend and treat depression. Yet, as recollections can be selected, influenced, or completely false as well as based on people's present perspective, *it can be countertherapeutic to use them to find origins of depression.*

While this knowledge may unnervingly oppose your prior understanding, fortunately, as memories can tell a revealing narrative they can be productively employed in treatment.

People's quest for awareness is admirable. Yet, sometimes it can be misdirected and taken down the wrong path. This leads you to investigate another widely-held conception that's a key to demystifying dysthymia.

Insight

When people want to know something (e.g., vacation resort location, a recipe, sexual health information) they often search the Internet. Or, when they want to see how things are done, they may look online for an instructive video.

Simply put, people regularly seek information and insight when they want to achieve an objective. Consider how this relates to depression.

Dysthymic individuals can be baffled as to why their symptoms exist and are curious about how to resolve them. With good intentions, they may go online to comprehend and soothe them. Yet, without the essential training and clinical practice, their comprehension and conclusions are likely to be incorrect and they may magnify their symptoms.

Nevertheless, this noble quest for knowledge should make perfect sense. After all, they're just employing a previously-productive investigative skill. Such free, easy, and anonymous Internet searches, for instance, often precede the next logical step: setting an appointment with a psychologist. (Which isn't necessarily free, or easy [as people can be apprehensive about making the first phone call], and certainly not anonymous.)

When you think about it, psychologists are highly-trained detectives of a sort, with a reservoir of information and expertise on the subject. Their skills are honed throughout their graduate training, professional development, and clinical experience. Similar to other experts, they're able to perceive clues and connections that are elusive to laypersons. Psychologists assemble an array of diverse puzzle pieces to form a coherent and penetrative picture of what's going on cognitively, emotionally, and behaviorally, as well as illustrate how people's goals and thinking are related to symptom development and persistence.

For example, people may not be aware of how they have unrealistic expectations of themselves or others that lead to symptom development. Or, they may not see how they act similarly to those they detest or disregard. In addition, individuals may not perceive when they employ excuses (which appear legitimate to them, but not to others) to withdraw from challenging situations or engage in behavior that's otherwise impermissible. (I assure you that all of the above is quite common, and such revelations can be staggering. Indeed, people are often stunned to the point of being momentarily speechless when their behavior fully dawns on them.)

Some individuals believe they unconsciously hold some way of thinking that if brought to the surface would not only explain their depression, but also illuminate a path to symptom reduction. (Odds are that you've seen this all-too-common plot device in movies and TV shows in which a lightbulb moment brings a character irreversible clarity and instant behavior change.) This belief can spur an energetic exploration of various unchartered cognitive territories to gather awareness they trust will swiftly deliver an anticipated benefit.

This seems valid, logical, and useful, but insight is insufficient for change.Just think about the following examples:

There are people who are well-aware they're overweight (or concerned family members or physicians repeatedly express their observations), but insight doesn't guarantee they'll change their eating patterns or exercise level.

Now, think about how some individuals are fully cognizant that they spend needlessly, have various bills due, regularly receive notices from collection agencies, and should reduce their spending. But this insight doesn't necessarily get them to adopt more prudent spending habits.

Cigarette packaging with clearly-displayed ominous warnings that the product can be dangerous – and the (sometimes rather gruesome) public service announcements which reinforce that knowledge – hasn't stopped everyone from smoking.

In fact, people often continue to drink alcohol to excess, use illegal drugs, drive recklessly, have affairs, gamble, and do other risky and self-injurious behavior while being fully cognizant, yet unconcerned, of the harmful effects. These examples should lead you to an unmistakable conclusion that can be applied to comprehending and treating depression.

Though insight can be helpful, it doesn't guarantee change in thought, emotion, action, or symptoms.

Complexity

As mentioned earlier, depression often brings with it various physical conditions; for example, fatigue, notable increases or decreases in appetite, an inability to sleep (or remain asleep) as well as oversleeping, reduced ability to concentrate, indecisiveness, decreased sexual desire, and despondency.[42]

Odds are, you can see how each of these symptoms can be related to depression...*as well as provoke or worsen it*.

In addition, medical and organic issues can mimic, coexist, or instigate depression in general, such as an acute cerebrovascular accident, meningiomas, acute myocardial infarction, non-steroidal anti-inflammatory medication, beta-blockers, hyponatremia, hypercalcemia, hyperthyroidism, adrenal insufficiency, Cushing's syndrome, Parkinson's disease, and Lyme disease.[43]

Decrypting dysthymic depression becomes even more complex and challenging after you factor in numerous other factors that might influence mood (e.g., financial straits, social rejection, relationship difficulties, career challenges, health concerns, lack of exercise, exasperating family interactions, sexual dysfunction, and alcohol intake – just to name a few).

Diagnosing diverse and varying factors as well as possible interaction effects can seem as overwhelming as trying to visually track specific birds in a zigzagging flock as they dart through densely-packed trees.

There are countless elements associated with neurotic depression, which makes it difficult to decipher.

So, consider what people often do when they want to understand, predict, and control complicated and intricate things.

Perhaps it's best to ease into this with an example.

You can't figure out how a plane engine works by looking at it in its entirety.

You have to delicately dissect it and examine how each part functions as well as how it's employed in relation to the whole.

So, through careful investigation of its vital organs, arteries of fuel delivery, oxygen and carbon monoxide respiratory structures, metal skeleton, electrical nervous system, as well as energy storage and combustion centers, you gather information about the way it operates, how it will run in the future, and become more accurate in predicting what might cause problems, as well as develop an increased ability to influence the way it works.

Reducing complex entities into smaller, more understandable elements can be an effective method to explain, predict, and control them.

In fact, the more convoluted and complicated the matter under study, the more difficult it is to grasp it completely. Therefore, it may be necessary to reduce what's being investigated to its most basic components.

This technique methodically eliminates as many extraneous variables as possible to get to the root cause or essential appreciation of some event or condition. A paint-by-number explanation of a complex portrait.

With that in mind, consider what fundamental elements scientists and researchers examine in an attempt to comprehend, explain, and treat depression.

Genetics and Being "Hardwired"

Dysthymic individuals are agonizingly familiar with how depressive symptoms can shatter their sense of peace, arrest their movement toward stated goals, as well as imperil relationships, socializing, and their careers.

Consequently, they may perceive their depression as an unpredictable and unstoppable bully who perpetually threatens their will, wants, and well-being. And, as human beings are a curious lot, they tend to wonder why they have symptoms – often in a virtuous attempt to forecast and remedy them.

So, consider what procedures they might employ to gather understanding, predictability, and determine which form of treatment to follow.

Often, they start gathering clues and collecting data.

For instance, depressed persons may look for associations between their symptoms and other factors (e.g., when they arise, who or what is nearby), or find parallels between their and others' symptoms.

Knowing that, think about where they might look first.

Just consider how frequently people identify genetically-based patterns among family members (e.g., height, eye color, hair color, skin tone, allergies). These things are easily observed and such associations have been made throughout history. Family members are close by, their appearance and actions are familiar, and modern science validates this timeless custom.

This practice is so long-standing and common that it's often rapidly and seamlessly applied to understanding depression. Fairly regularly, dysthymic individuals will reference family members who demonstrate similar symptoms and quickly conclude that their symptoms are genetic in origin. And, given people's proneness to look for patterns, there's also a tendency for *others* to make observations such as, "You got your father's genes. He's been depressed his whole life" or "You inherited your mother's moodiness." This tendency appears logical and potentially productive. Moreover, it can support the perception of a genetic basis of dysthymic depression.

However, there are associated difficulties.

First, people are prone to seeing and seeking patterns in significant, as well as worthless, information.[44] (The perception of a non-existent relationship is known as an illusory correlation.[45])

For example, what do the following suggest?

"Our whole family loves honky-tonk country music."

"You have your father's gene for working on old boats."

"You inherited your mother's fondness for online shopping."

The above illustrate *learned* characteristics.

As it's cognitively easier and familiar, people tend to perceive a genetic cause – *even when there isn't one*. (Amusingly, I once knew a woman who looked as if she were her mother's clone. The statement of a genetic link was nearly-instinctive…but a revelation quickly followed: The daughter was adopted.)

Now, take a moment and pick any one of your non-genetically-based habits, say watching football, playing video games, or what type of books you enjoy. Next, is there someone in your extended family that also enjoys those things?

Simply, if the number of people or quantity of characteristics is sufficiently large, odds are that you'll eventually be able to see a pattern or connection. Taking this a step further, you can see how the genetic link and the enjoyment (or hatred) of a specific activity, may be erroneously paired and give the false impression that there's a gene for watching football (for instance) which you and the other family member share.

Dysthymic persons may identify a depressed family member (and, perhaps revealingly, overlook the majority of non-symptomatic ones) and presume a genetic link and cause of depression.

So, if two (or more) genetically-related family members are depressed, that doesn't guarantee a genetic origin. There can be a misleading belief of a genetic basis when people don't take other factors into account (e.g., role-modelled thought and behavior patterns, what's socially allowed).

Second, when dysthymic persons (or others) perceive parallels in thought, action, and emotions between them and their relatives and attribute those similarities to genetics they tend to arrive at certain – and perhaps dismal – conclusions, which may perpetuate symptoms.

Now, consider what this can look like.

For example, people may tell those with dysthymia, "Your brain is hardwired for depression. It'll never change. You'll have to take medication for the rest of your life." This portrays wrecking ball-like symptoms as stubborn, permanent residents, unable to be evicted.

Consequently, dysthymic persons may feel hopeless and helpless.

Third, when people testify their symptoms have been there for years, or for as long as they can remember, consider what they might imply.

When individuals perceive their symptoms as present since childhood, they tend to believe (and suggest to others) that their depression was innately inscribed in their genes when they were developing in the womb.

In fact, I've talked to many people who resolutely state that they're "hardwired" for depression. They conclude – often not based on any medical examination of their DNA – that their depression is inborn and irrepressible. Consequently, some think they're in an unwinnable situation, and become too demotivated to enter the ring and fight their symptoms, with the firm belief it would lead to an inescapably bloody and frustrating forfeit.

Consider how a purely genetic origin of dysthymic depression could readily explain why it often appears alien and counter to people's goals. Yet, if it were only genetically based, then you must face a number of questions:

- Wouldn't medical research provide more clarity and certainty in detecting those prone to developing dysthymia (as they can for breast cancer, Alzheimer's disease, Parkinson's disease, and Celiac disease)?
- Might medical science be better able to control it?

- How do people develop symptoms when they *don't* have a family history of depression?
- Various epigenetic (external, non-genetic) factors can turn genes on or off.[46] However, does this account for the wide range of depressive symptoms (e.g., which ones arise, when, where, with whom, if alone) among various people or within an individual?
- Why do symptoms erupt or evaporate (hint: often routinely) when there's a change in non-genetic factors (e.g., social or physical setting, in the presence or absence of certain challenges, such as work or dating)?
- Given that depression can sabotage intimacy, relationships, and cooperation, for example, might natural selection have eroded the prevalence or potency of dysthymia within the population?
- How do you explain those who resolve symptoms through talk therapy?

Although genetic causes, or predispositions, to depression exist, they can have minimal influence, be vague, related to several diagnoses, open to interpretation, and be subject to environmental influences.[47]

As genes can be turned on and off, even if people are genetically predisposed to depression that doesn't mean those genes are expressed – *genetics aren't the only factor as to whether symptoms arise.*[48]

But there's something else to consider.

Given the variation of physical compositions and physiological functioning within the brain (among other organs), it's simply impossible to state that there's only one cause, reason, or presentation of depression. (If a parallel will be helpful, consider how there are many reasons why you might get hungry and start eating, e.g., you're actually hungry, other people are eating, you're bored, someone brings food you've never tried before, you're avoiding some other task, it's tradition, you don't want to seem odd being the only person not eating, or you eat different types of food depending on what you find appetizing or what restaurants are open or what leftovers are left in the fridge).

While research shows that there are some hardwired conditions and responses – such as reflexes – given the neural plasticity of the brain, people are quite flexible in their psychological and behavioral expressions, which environmental experiences can shape.[49]

This includes depression.

The practice of reducing complex entities to simpler elements is a common, and often productive, practice. Nevertheless, it appears that reducing the complexity of dysthymic depression down to only a purely genetic basis can be insufficient and potentially problematic.

Knowing this, consider how else people reduce depression in an attempt to understand, predict, and treat it.

Chemical Imbalance

If you've ever had a cold or flu, you have a glimpse of how microscopic factors can poison people's thoughts, attitude, physical appearance, and enthusiasm, as well as spoil and oppressively limit their ability to do vital actions, such as eat, work, socialize, and have sex.

Or, think about how excessive alcohol consumption can lead to slurred speech, impaired judgment, headaches, reduced attention, memory issues, nausea, dizziness, distorted vision and hearing, drowsiness, shakiness, sweating, slow reflexes, numbness, increased urination, vomiting, sexual dysfunction, and black outs.[50]

Now, consider what these examples illustrate.

Simply, it's that biological or chemical disharmony can detrimentally impact people's mood, cognitive clarity, appearance, behavior, and energy level – perhaps to the point of imperiling their ability to survive and thrive.

Given the widespread knowledge that a virus, chemical, or some other tiny organic influence can create disproportionally-immense symptoms, consider how this might guide people's comprehension and treatment of depression.

To start, acknowledge that physical and chemical issues – for instance, those associated with brain injury or dementia – can significantly change people's cognition and behavior, and may produce depressive symptoms.[51]

And, in fact, there are studies which show that depressed persons can have different neuronal activity and chemical levels in particular brain regions from non-depressed people.[52]

So, it's understandable that some dysthymic individuals earnestly state that a chemical imbalance causes their symptoms. Many depressed people go one step further and identify and accuse the neurotransmitters dopamine and serotonin as the guilty culprits.

Consider how this mindset may influence people's treatment path.

As an example, think about those who eat serotonin-rich food with the firm belief it will balance the amount of chemicals in their brain and reduce symptoms. Often, they feel noticeably better after consuming particular snacks or meals. This result reinforces their perception of a chemical imbalance as well as the use of a seemingly simple, natural, and effectively-accurate solution.

Now think about the more scientific treatment.

Virtuous and dedicated doctors, chemists, and researchers formulate psychotropic medications to influence people's mental state.

Antidepressant medications reduce symptoms and using them in conjunction with psychotherapy may magnify the positive results.[53]

People generally prefer easy-to-remember rules, concise answers, and straightforward solutions. The concept of a chemical imbalance and taking a pill fit those criteria. It's as familiar, simple – and sometimes nearly as quick in reducing symptoms – as using an antibiotic for a cold.

Pharmaceutical advertisements may fortify and hone the perception of a chemical imbalance and (unsurprisingly) prompt the sale of medication to fix the perceived problem.[54]

Now think about how after people medically reduce their symptoms they may…

- Fortify their belief that a chemical imbalance caused their depression.
- Testify to the benefit of psychotropic medication.
- Say they're content to take antidepressants for the rest of their life, if necessary, rather than return to how things were – fearing a horrific sequel of symptoms if they were to discontinue taking their medication.

The mindset that a chemical imbalance causes depression appears to easily resolve the mystery of dysthymic depression.
1. It seems to provide a strong foundation on which to build a robust and readily available set of medical remedies.
2. The use of antidepressant medications looks to be a rational, appropriate, and productive treatment.

3. A chemical imbalance perspective might avoid the assessment and therapeutic limitations associated with insight and memory retrieval, for instance.

So, in light of the clinical research and personal testimony (as well as countless commercials), it appears as though there's clear and irrefutable proof that chemical imbalances cause depression. Therefore, you can see how people would logically use antidepressant medication to efficiently and effectively alleviate their symptoms.

Well, there you go. Everything seems to add up, mystery solved, a clear and uncomplicated conception and remedy for depression…wait, no, no, hold on, this is where things start to get a bit more curious, complicated, and cloudy…

Chemical Imbalance Questions

Take a moment to think about those who believe that a chemical imbalance of serotonin in their brain causes their depression and how they experience symptom reduction after eating select items. They appear to employ a simple, organic, and effective remedy.

However, eating serotonin-rich food *doesn't* influence brain serotonin levels, as serotonin cannot pass the blood brain barrier.[55]

So, as noble and natural such a home-brewed solution is, it's inaccurate. But if that's the case, consider why people feel less symptomatic after eating food they believe to be medicinal.

There's a number of reasons. For instance, it could be the placebo effect, or experiencing a reassuring sense of control over their seemingly unpredictable and unassailable symptoms, or merely because they're directing their attention away from depressing things and onto desirable, but otherwise dietary-restricted, food – such as chocolate.

So, that may not work as believed, but consider scientific studies.

You may say, "Just a couple of pages ago *research provided clear evidence* that depressed people have different levels of neurotransmitters and neuronal activity than non-depressed persons. Doesn't this prove that a chemical imbalance causes depression?"

That seems reasonable, but recognize some confounding variables.

It may help to start by imagining automatically getting up from the couch and heading toward the door on Halloween after hearing a doorbell…from a show you were watching on TV.

Now, recall how confirmation bias defines people's tendency to acknowledge those occurrences and conditions for which they're primed.

When people are instructed that a chemical imbalance causes symptoms, they may be increasingly prone to look for information and experiences which confirm that perspective.

So, when dysthymic persons hear that certain neurotransmitter and neuronal activity levels coincide with depression, they may default to interpreting those findings as proof that specific microscopic factors cause their depression. Unquestioned, this impression may be accepted as self-evident and universal.

Ah, but there's something else you must think about.

Perhaps an analogy can help.

What if someone told you that fire trucks cause house fires? After all, they say that you just need to look at the evidence.

1. Wherever there's a house on fire, at least one fire truck is present.
2. There's an exceedingly low probability that there are fire trucks at houses that *aren't* on fire.

Even though you *know* that fire trucks don't cause house fires, there's a strong correlation that risks sustaining a misperception. *A correlation between two conditions doesn't prove causation.*[56] Or, if you prefer Latin, "cum hoc ergo propter hoc ('with this, therefore because of this')."[57]

So, knowing that two conditions co-exist cannot be used to prove causation, some curiosities are stirred.

First, there's a question of whether there's any *association*. For example, the Big Ben in London co-exists with the planet Jupiter, but there's no connection. Although there are physical and chemical conditions that can cause depression, is that universally applicable to all forms of depressive symptoms?

Second, there's a question of *directionality*.

Entertainingly, there's what's known as the Hemline index, it's a correlation between the stock market and hemlines; when the stock market rises, dress hemlines suggestively ascend, and when the stock market goes lower, hemlines go down – but, the hemlines do not create, predict, or precede the movement in the stock market.[58] There's an association and identifiable directionality. Yet, people can misperceive the direction of influence, believing that hemlines influence the stock market.

Undeniably, physical or chemical conditions can pre-exist and create depressive symptoms. This testifies to an association and direction.

But also think about it the other way around.

Focusing on depressing events and conditions may influence neuronal activity and neurotransmitter levels.

Sure, there are times in which that happens.[59] This would explain, at least in part, how thinking can lead to observed differences in neuronal conduction and biochemical levels. (If in doubt, just think about how people watching their favorite team win an important game can influence their heart rate, happiness, rate of speech, physical activity, etc.)

Accordingly, if thinking can generate characteristic neuronal activity and neurotransmitter levels associated with depression, consider how changing the way people think may change those factors and reduce symptoms.

Indeed, psychotherapy can influence people's biochemistry and neuronal functioning.[60] Just like updating your phone's software impacts how the hardware works (e.g., quicker screen refresh rate, longer battery life), or how parents can soothe their visibly distressed kids by redirecting their attention or reframing an experience, psychotherapy treats people's way of thinking which may persuade neuronal activity and neurotransmitter levels.

Third, if there's a correlation between depression and identifiable neuronal activity and neurotransmitter levels, *and* you're aware that there can be a causal link, *and* you know that there can be differences in directionality, then you face another question: *What came first?* The answer may illuminate depression's shadowy, uncertain corners. Did certain neuronal activity and neurotransmitter levels pre-date depressive thoughts and perspectives, or vice versa?

Indeed, it can be difficult to decide if one precedes or causes another, or occurs simultaneously (related or not).

As an indirect illustration, about a third of dysthymic individuals have a history of substance abuse, but it isn't clear whether drug use leads to dysthymic depression *or* dysthymia leads to drug use, or another factor (or more) must be taken into consideration.[61]

Seeing a correlation can risk a projection of what came first, which in turn may influence people's assessment and treatment choices.

Fourth, whether specific chemical or neuronal activity levels foster depression, or depressive thoughts cultivate symptoms and related neurotransmitter and neuron activity levels, you encounter a new question: *What's the degree of influence?* That is, even with a cause and effect condition, there may be uncertainty of *how much* authority one has over the other.

When examining chemical imbalance causes of depression, you must consider association, directionality, what came first, and the degree of influence.

Believe it or not, there's yet another confounding variable to contemplate.

Think about how during nearly everyone's childhood, medication treated microscopically-based physical symptoms (e.g., achiness, nausea, headaches, rashes). With that in mind, note that neurotic depression can have noticeable physical manifestations (e.g., low energy, poor appetite, sleep issues).

When people develop – especially early in life – a template that physical symptoms have a microscopic physical cause and are treated with medication, consider how they might perceive *psychologically-*based physical symptoms.

Perhaps in accord with their view that physical symptoms have a physical cause and are treated medically. The notion of a chemical imbalance causing depression and the need for medication fit the pre-existing conception as neatly as test tubes in a centrifuge.

There's one more factor that might magnify people's belief that a chemical imbalance causes depression.

To start, consider how well-advertised movies and music may occlude and misdirect people's attention from lesser-known films and songs that would be more suitable and they would enjoy more. (Indeed, how often have overhyped summer movies disappointed you?)

It's rather common that those with persistent depressive disorder confidently state a chemical imbalance is directly responsible for their symptoms. Perhaps revealingly, this often occurs when they have *no personal medical data* that can confirm or deny this claim.

When dysthymic persons (and perhaps their concerned friends and family) are in the dark about other factors related to developing depression, they may inadvertently multiply their belief in a spotlighted chemical imbalance cause. Maybe more so when those other factors are complex, convoluted, and subtle, and the related treatment is challenging, intricate, and indistinct.

(As an intriguing aside, the United States and New Zealand are the only two countries that allow direct-to-consumer advertising of *prescription* medication [rather than permitted to only market directly to trained professionals who prescribe drugs].[62])

Dysthymic persons may misperceive a psychologically-based depression as due to a chemical imbalance and in need of a medical remedy. When antidepressants reduce symptoms, people can inadvertently fortify their misbelief of a chemical imbalance and reduce the likelihood they'll investigate other conceptions of symptom development and forms of treatment.

When people talk about a chemical imbalance, they often refer to dopamine and serotonin levels as being directly responsible for their depression.

However, it isn't as clear cut as that.

Think about how electricity runs throughout your home. It's *generic* in character, as you use it for various purposes, say to power a TV, refrigerator, furnace, bathroom fan, or microwave oven. The same thing can be said about water. It too runs throughout your home and you use it in a variety of ways, for example, to shower, cook, clean clothes, or water plants.

With that in mind, note that dopamine and serotonin are employed in numerous ways throughout the body, and their effect can depend on where they're released (e.g., Parkinson's disease, dementia, growth hormone secretion, reward, memory, aggression, schizophrenia, major depressive disorder, decision making, sleep, sexual behavior).[63]

Given their general and widespread use, dopamine and serotonin may not be as specific or exclusive to depression as some people believe.

How the brain and the body work can be as complicated, indirect, subtle, and mysterious as covert military operations. Also, there are many interacting variables within an individual, such as biological functioning and genetic recipe, as well as various psychological, personal, social, diet, financial, activity level, and occupational factors. In addition, diversity among people may multiply the variation in symptom development, assessment, and treatment. This may explain different individuals' diverse responses to medication, for example.

Interestingly, there's some debate about the notion of a chemical imbalance, as it isn't known whether there's an optimum level of chemicals in the brain.[64] Furthermore, there's research that suggests that a chemical imbalance doesn't have strong supportive evidence.[65]

Now, consider what might this mean.

If it's unclear what constitutes the healthy level of neurotransmitters, then it's difficult to define an imbalance. Think about it in this way. If you knew that someone's weight was 175 pounds, you wouldn't know if that's healthy or unhealthy if you don't know other relevant factors (e.g., the person's height, sex, muscle mass/fat ratio).

Simply, it's difficult to say that only one factor is either responsible for – or a resolution to – depressive symptoms.

Debate within scientific literature and practices is prevalent and well-established – and done for good reason. Consequently, you needn't dismiss biology, chemistry, physiology as unrelated to symptoms. Nor, should you view medication as useless, or medical doctors and researchers as ill-informed or malicious. In fact, medication can reduce symptoms, and be incredibly useful for those who must address their responsibilities in life, as well as able to assist people's progress through psychotherapy.

As there's variation in understanding and treating depression, the cause and effect-based mindset of a chemical imbalance may be insufficient for full comprehension, and not universally applicable.

This isn't to rule out, minimize, or criticize existing concepts or treatment, but rather beneficially allow you hints and glimpses of other ways to perceive, assess, and treat persistent depression. Having a wide perspective of clues and evidence may reduce the risk of following a single, alluring, but serpentine distraction that leads to you a false conclusion while unraveling this mystery.

Now consider what practice the cause and effect mindset can demand that may influence how people perceive and approach their symptoms.

Searching for a Cause

People have a remarkable and advantageous ability to perceive patterns. This skill is often applied to assessing situations and observing a link between two or more conditions. For instance, identifying a connection between eating a particular food and being sick enables the avoidance of future illness, or developing a rash after touching a plant with jagged, three-pointed leaves can lead to increased vigilance of those distinct characteristics to avoid repeating that unpleasant experience.

Indeed, the capacity to detect relationships is essential to learn the vital construct of cause and effect, which can assist people's pursuit of a safe and healthy way of life, as well as grasp how things work as a means toward goal achievement (e.g., knowing what tools and software applications can allow people to accomplish), and is indispensable to survive and thrive.

Just consider how married individuals can be well-versed in what words and actions their spouse hates, as well as what sweet statements and kind deeds are associated with attentive and rewarding sex.

Note that the preceding chapters focused on searching for causes of depression. In other words, identifying what patterns people perceive between an event or condition and symptom development.

This mindset guides comprehension, assessment, and treatment, from its earliest stages to the present. For example, immorality was once deemed as creating symptoms, whereas more current concepts and practices have people searching their memories for a reason why they're sad, or developing insight about what triggers their depression, to examining what role genetics play in producing symptoms, whether people have hardwired causes for depression, or that a chemical imbalance is a catalyst for unrelenting misery.

Though some of the later practices can provide clinically useful information, all share a cause and effect mindset, which may narrow a wider view of depression, perhaps limiting comprehension and care.

But consider how searching for what causes depression may be misleading.

Recall from earlier that perceiving non-existent relationships is called an illusory correlation. Relatedly, patternicity and apophenia are tendencies for people to see patterns in meaningful as well as meaningless data.[66] For instance, people perceive entertaining and innocent images in random, indistinct things (e.g., clouds, woodgrain), or they can be unhealthy (e.g., superstitions).[67]

Notably, this can lead to various outcomes, from being amused and amazed (e.g., laughable misperceptions, optical illusions) to problematic (e.g., believing that people are truthful when they're not, stereotypes) to lethal (e.g., identifying the wrong person at a criminal trial, misdiagnosing a serious medical condition as one that's harmless).

People may overuse their ability to perceive patterns and relationships.

The tendency to find incorrect or irrelevant relationships or patterns is a side effect of people's need to identify cause and effect relationships – as well as other useful patterns (e.g., discern the face of a predator from nature's canvas).

In addition, when people search for a cause how think about how they can fall prey to inattentional blindness (an inability to see one thing because they were determinedly looking for something else).[68]

Well, think about how operating from within a cause and effect paradigm depressed individuals – and perhaps their therapist – believe that some event or condition produces symptoms. Consequently, they must satisfy their hunger for a cause. This is often accomplished by sleuthing through current and childhood recollections for any negative or traumatic event in an effort to grant clarity and guide treatment. Yet, as you know, memories are often inexact and prone to various errors, misguiding influences, and fabrication.

The cause and effect perspective can prime people to look for a source of symptoms, and inattentional blindness may distract them from investigating other elements and factors that can explain depression.

Also, as everybody has an abundance of sweet and bitter experiences throughout life, it's ground so fertile that if necessary anyone could pick some unsavory event or condition as the basis of depression.

The belief that there's a cause may not only lead to a search for one, but may guarantee finding one, whether or not it's accurate, and make it less likely to look elsewhere for understanding.

Now, consider the following.

1. The cause and effect mindset prompts a search for symptom causes.
2. People's tendency to perceive patterns can be overused.
3. They may perceive non-existent patterns, say misidentifying something as a cause (e.g., childhood experience, genetics, chemical imbalance).
4. In the quest to find a cause, inattentional blindness may prohibit people from detecting other factors related to symptom development that might counterbalance the limitations of the cause and effect perspective.

This *isn't* to say that biological, chemical, genetic, or physiological origins of depression should be shelved as fiction. Rather, could the cause and effect narrative of dysthymia have certain plot holes, contradictions, and distractions that may prime people to perceive a cause of depression *even if it doesn't exist?*

With that, think about what people can acknowledge that may broaden their perception, understanding, assessment, and treatment of depression.

Symptom Functions

Dysthymic depression can be painfully debilitating. Sufferers can feel dreadfully inferior, awkward, imprisoned, and alone. Moreover, symptoms may arise unpredictably with seemingly no reason, prohibit vital functioning, be difficult for nearby others, as well as sabotage relationships or employment.

Consequently, persistent depressive disorder can appear innately generated (e.g., genetically, biologically, chemically, physiologically) and only treatable by medical means. Or, people may deem experiences (e.g., childhood events, grim financial situation, rejection) as the cause. Yet, sometimes the cause and effect mindset of depression can fall a bit short…but the mystery doesn't have to go unresolved. There's another avenue to understanding.

Just think about how *biological* symptoms can help you detect what's missing in comprehending *psychological* symptoms.

For instance, imagine a sweltering summer day when those outdoors perspire because their body temperature rises beyond an optimum level. From within the cause and effect framework, it's easy to see how heat causes perspiration.

But stopping your understanding at that point can be an injustice to the wonders of the human body.

While perspiration can be seen as a symptom of being too warm, as the sweat evaporates it carries away heat and reduces body temperature in the process.[69] It's a practical and impressive system, using that which is causing a problem to be part of the solution. (It's also interesting, and perhaps slightly entertaining, to note how any malodorous manifestations may keep other warm bodies that can worsen the problem coolly distant).

If you desire another hint, think about how the body responds after being cut.

To stop blood loss and promote healing, the body redirects blood flow and platelets race to the wound to promote clotting. White blood cells speed to the injury and eliminate harmful bacteria to halt or reduce the likelihood of infection. Dead cells and tissue are removed. The wounded area may contract to make the healing process more efficient and effective. New cells form, tissue grows, and cells are repaired to restore previous structures and functions. Final cleanup of the wound is achieved through the formation of scar tissue and its eventual reduction or removal.[70]

Amazing! Yet, such fascinating and useful orchestration of various bodily systems isn't an anomaly.

In fact, consider what the following reveal about symptoms.

- The gag response may prevent choking and asphyxiation.[71]
- Fever elevates body temperature, which promotes sleep and aids immune functioning.[72]
- Symptoms are means to perceive appetite, starvation, and malnourishment (e.g., hunger pangs) and provoke eating.[73]
- There are innate mechanisms that detect thirst and dehydration (e.g., discomfort, dry mouth) and prompt re-hydration.[74]
- Pain is associated with dangerous circumstances such as, problems with dentition, lacerations, skeletal fractures, and drowning.
- Vomiting, mucus, pus, and diarrhea can minimize damage from viruses and bacteria by evacuating harmful substances, as well as promote recuperation.[75]

Consider how these functions occur automatically (without cognition) and through the cooperation of assorted bodily systems, actions, and chemicals. This provides illuminating evidence of how the body is genetically programmed to engage in specific coordinated action to achieve the goals of survival and thriving (though often not in particularly attractive or comfortable ways).

This leads you to an unmistakable and indisputable conclusion.

Bodily symptoms can serve various purposes.

Next, consider how the agony associated with a broken leg immobilizes people, which can prevent further damage. Or, pain can prompt action to move away from leaning against or stepping on something sharp.

Such responses highlight and assist symptom purposefulness.

Symptoms can helpfully influence an individual's movement.

"Wait," you might insist, "isn't that only a reflex?"

Well, it can appear that way initially, but symptoms are complex (perhaps far greater than you've given them credit).

Now, consider what's revealed when someone painfully catches on fire and runs to a garden hose or fire extinguisher, or drops to the floor and rolls around.

Such responses seem like a simple, straightforward instinct. However, upon further investigation you may detect some enlightening clues.

1. The symptom of pain (as well as the observation of being on fire) acts as a feedback mechanism, immediately highlighting the urgency of action, prompting the person to prioritize precise goal selection: *Exterminate the fire and the related agony as soon as possible!*
2. The person instantly *chooses* what's required to achieve those goals. (And think about the evidence for this.)
3. The individual rapidly initiates symptom-fueled action which has a particular direction toward a certain item or place (i.e., a garden hose, fire extinguisher, the floor) in an attempt to achieve the chosen objectives.
4. The specific symptom-fueled movement *stops* after attaining the goals (e.g., the person is no longer on fire).
5. Symptoms can prompt the selection and achievement of *new* goals and associated action (e.g., apply burn relief ointment, run the scorched area under cold water, go to the hospital for care).

This example illustrates some important facts.
- Symptoms can act as a feedback mechanism that identifies a threat to survival and thriving.

- Symptoms can spark the instant selection of new goals (e.g., from casually talking while making dinner to feverishly putting out a fire).
- Symptoms can prompt rapid or immediate action that's in accord with the chosen goals.
- Symptom-enabled cognition and movement can be purposeful – though not necessarily a fully-conscious process.
- Symptom-fueled movement can reveal people's preferences and thinking (e.g., goal selection of putting out the fire [versus trying to show off], rules of action are shown in the chosen method of how to extinguish the fire [running to a garden hose versus going to someone else to have that person put out the fire]).
- Symptom activation, cognition (e.g., choosing objectives, as well as the evaluation and decision of appropriate movement), and the related behavior can all happen with tremendous speed.
- Identical symptoms (e.g., pain) can be used to guide different behavior depending on people's goals (e.g., First, to extinguish fire. Second, to cool burned areas with icy water. Third, to go to the hospital).
- Symptom-prompted goal selection and associated action may appear to be a cognition-less reflex, yet symptom purposefulness and thought process is further revealed and validated once you recognize that symptom-fueled actions that *aren't* in accord with the goals do *not* occur; for instance, people choose and move toward a fire extinguisher rather than a gas can, grandfather clock, ham sandwich, or tuba.

But, there's something else to keep in mind.

Consider what would happen if you were to rest your hand on a hot stove without the ability to experience pain.

You'd instantly realize that pain can be a protective signal. In fact, some people are born with a congenital inability to sense pain (CIPA), who suffer injuries and accidents (e.g., broken bones, biting through their tongue, burns) that would have been avoided if they had the ability to detect pain.[76]

Pain is one of the body's innate feedback mechanisms. It can communicate a problem – the loud, obnoxious friend who not only bluntly tells you when you've crossed the line and what to do about it…but continues to annoyingly rub your face in it for some time afterward.

Indeed, pain can teach a lasting lesson, recalled at will.

In this way, you can view symptoms not only as a feedback mechanism that guides present thoughts and actions, *but also shapes future cognition and behavior* (e.g., Stoves are hot; don't touch them again!).

While physical symptoms can appear to be purely a physical matter, it's necessary to keep the psychological piece in mind.

Just consider how people can experience pain when they pick up something that's far too heavy. That symptom reveals things about their thinking.

1. They thought the object is lighter than it actually is.
2. They believed they were stronger than they truly are.
3. Their strategy for lifting heavy objects is inadequate.
4. All, or a combination, of the above.

Thoughts can precede behavior, even if they're short, incomplete, automatic, or even below conscious awareness. This leads to a particular conclusion.

There are times when physical pain can be a feedback mechanism which communicates to people that their perception and way of thinking is incorrect.

As people may be notably reluctant to change their way of thinking, they may stubbornly refuse to heed their body's feedback. Unfortunately, such willful disregard can worsen their symptoms e.g., an immobilizing wrenched back can result when people do not listen to the initial pain message when lifting an object. (And consider how this parallels what some dysthymic persons do.)

Knowing this, think about how *mild* pain messages might be useful.

Consider how slight discomfort when eating can be a signal to put the fork down and walk away from the table. Or, how moderate bladder pain can be a gentle message to discontinue whatever you're doing and head for the nearest restroom, otherwise regrettable events may occur. Among other things, mild pain messages can direct people to stop behavior *before* it becomes problematic.

Evidently, even minor pain messages can be advantageous.

Now, consider what the benefit of having a range of pain.

Think about how a heart attack is more painful – and should be – than bumping your arm. (Though exceptions exist, e.g., unexpected papercuts and shocks from static electricity can be disproportionally painful. Also, there are serious conditions that don't generate much, if any, pain, e.g., brain tumor.)

Being able to perceive a range of pain – from mild to severe – can communicate the degree of damage or concern as well as enable people to appropriately prioritize their action for addressing the issue.

Pain can dutifully warn people of invisible, internal problems that need medical care (e.g., cancer, slipped disc, kidney stones, sexually transmitted infection, explosive distress from food that should've been avoided).

Keeping with the pain example as a simple, familiar, and generalizable illustration, recognize that no one else can truly feel another person's pain – it's a symptom that exists solely within the individual.

Yet, there can be a social component to it.

Just think about how sharp pain from a broken bone, for instance, can trigger yelps that alert others of an emergency and appeal for assistance. Or, consider how new parents rush to their colicky newborn's cries.

Symptoms can have a social factor as they can influence others' movement that benefits the symptomatic person (and perhaps the nearby persons as well).

Of course, the social aspect of an individual's physical symptoms extends beyond pain. Various symptoms can communicate a medical concern to others, even when the symptomatic person is unaware of the symptom. For example, consider how an individual's pale and frail appearance, slowed or slurred speech, or wandering thought patterns displayed in frequent mid-sentence change of subject, notifies others of an illness and need for medical attention.

An individual's physical symptoms can arise instantaneously, consciously or unconsciously, be intricately and intriguingly choreographed, have a social component, and usefully employed to achieve various goals, perhaps most prominently to aid one's ability to survive and thrive.

This discussion of physical symptoms relates to how people perceive and treat psychological ones.

Symptom Suppression

Psychiatrists, primary care physicians, and non-physician providers (e.g., nurse practitioners) are professional, honorable individuals who prescribe antidepressants to reduce symptoms. And, quite impressively, psychotropic medications can humanely – and relatively quickly – liberate people from a distressing and limiting state. While some people can experience side effects of antidepressant use, a combination of drugs may be prescribed to alleviate them.

In addition, antidepressants may address symptoms regardless of their origin or association (e.g., neuronal, chemical, psychological, social concerns), and whether they're related to long-term issues (e.g., neurological problems due to a stroke) or shorter-term stressors (e.g., divorce, bereavement).

For these and other reasons, prescribing and taking antidepressants can be logical, compassionate, and therapeutic.

In fact, antidepressants are one of the three most commonly prescribed drug classes in the United States.[77] In a report published in 2017, nearly 13% of Americans aged 12 and over acknowledged using antidepressants within the previous month – an increase from a prior data collection study.[78]

But this trend isn't only in America.

While antidepressant use isn't uniformly on the rise, it has also increased in Iceland, Australia, Portugal, and the United Kingdom, for example.[79] This may suggest greater access to healthcare as well as psychotropic medication acceptance and effectiveness. Yet, people can be unclear about how they work.

Antibiotics, Painkillers, and E-mail Filters

How might people view antidepressants that rapidly reduce their symptoms?

To ease into the answer, examine a different medicine: Antibiotics.

During childhood, odds are you had strep throat or a sinus ailment.

It's also likely you visited a medical doctor, were given a prescription for antibiotics and took them, which quickly killed the offending bacteria. Once you conquered the infection, there were no more symptoms, you felt fine, and you stopped taking antibiotics given that you no longer needed them.

Such experiences may create a mindset that all medications are curative.

Indeed, many medications can resolve medical concerns (e.g., antifungals, antidotes for poisons). Using medication to cure illness is a nearly universal and typically positive practice that's socially encouraged, starts in infancy, and can get repeatedly reinforced throughout the lifespan.

Yet, it can be inadequate to only view medications as curative.

Just consider how your e-mail filter blocks irrelevant and inappropriate spam messages or other unwanted e-mails (say from insufferable co-workers or know-it-all relatives) from going into your inbox.

As you never see those unpleasant messages, you avoid experiencing any disagreeable emotions (or become tempted to respond in a less-than-civil manner). As long as your e-mail filter remains on, you won't see any of those relentlessly-sent unwanted e-mails in the future. But if you turn off that filter, those annoying messages quickly flood your inbox.

With this in mind, think about how pain can be the frustrating roadblock that arrests people's movement toward their desires (e.g., working, exercising, socializing). Appropriately, they take pain-alleviating medications (aka analgesics, painkillers). So, like parents who use serenity-protecting earplugs when accompanying their child at a deafeningly-loud concert filled with shrieking teenage fans, painkillers reduce people's physical distress so they can peaceably achieve their goals, say play sports, sleep, or go for a walk.

People often use analgesic medications to make aches and illnesses more tolerable while the body heals itself or until receiving proper medical treatment. Also, analgesics are vitally useful for excruciating medical conditions and surgery, as they soothe, silence, and suppress physical symptoms without influencing the underlying cause or interfering with healing.

Whether for major or mild pain, using painkilling medications appears undeniably reasonable, beneficial, compassionate, and practical.

However, painkillers aren't a cure.

Comparable to how an e-mail filter stops incessantly sent unwanted e-mails from entering your inbox, analgesics halt continually sent pain signals from reaching people's consciousness inbox, so to speak. As they're unaware of the pain messages, they don't feel discomfort.

Though sometimes they can reduce or resolve inflammation (e.g., ibuprofen), in general, pain-relieving medications don't stop the injury or cure an illness, but merely suppress the ceaselessly sent pain messages.

Knowing this, consider when using analgesic medications can be illogical, countertherapeutic, and irresponsible.

Just imagine if people perceive themselves as healed because they feel better and act in accord with that misperception. They may soon run into trouble (e.g., walk on a broken leg).

Or, consider how some people medicate symptoms *rather than address why they exist*. For instance, they take painkillers to avoid getting unpleasant news from a doctor (e.g., terrifying diagnosis, need for immediate surgery, the demand to improve their diet and exercise).

Physical symptoms (e.g., pain, vomiting, diarrhea) can indicate a problem as well as guide thought and behavior that assist survival and thriving. Yet, like an obnoxious neighbor who won't change or move, symptoms can persistently and infuriatingly lessen people's quality of life. So, to address their responsibilities and interests, it's logical that people use pharmaceuticals to silence discomfort, cure a condition, enhance their body's ability to heal, and minimize other concerns. When medication cures a problem, there's no symptoms to extinguish, so people stop taking the medicine as it's no longer needed.

Note that there's a difference between the curable and that which is manageable by medication. With that in mind, consider a potential risk that prescribers try to avoid.

Sometimes medically suppressing symptoms can reduce the efficiency and effectiveness of the body's healing and protective feedback mechanisms.

Medically interrupting, weakening, and muting innate symptom machinery may create a dangerously-attractive mirage that people are in good health when they're not, and inadvertently worsen their situation when it enables them to avoid healthier actions for an extended period of time.

This discussion about antibiotics, painkillers, and e-mail filters may allow better understanding of antidepressant medication.

Telling Analogies

Antidepressants can reduce people's symptoms that are related to diverse sources and situations as well as permit them to productively get on with their lives. That's terrific. However, there's something of which you should be aware…*or perhaps have already figured out*.

While antibiotics cure an illness by eliminating the cause, you can see how antidepressants may act more like painkillers or e-mail filters.

Turning off an e-mail filter permits junk mail to enter your inbox. Stopping analgesic medication can allow pain messages to be received. With that in mind, consider what may happen if people stop taking antidepressants when the conditions associated with their depression don't change.

Symptoms reappear.

Antidepressants may merely suppress symptoms that wait to resurface.

By and large, dysthymic individuals are fully aware that their psychotropic medications do not cure their symptoms, but rather anesthetize them (which still can be pleasing and useful, like a refreshing dip in a cool pool to tolerate a scorching day). While chronic symptom sedation may seem like an agitating lifelong sentence, think about when people might be at peace with it.

Often, it's when they believe their depression is genetic, biochemical, or neuronal in origin and requires medication to address the symptom. If this sounds odd, think of it as similar to how people happily correct other persistent innate issues (e.g., wear glasses, use hearing aids, take vitamins).

Consider when people may mistakenly view antidepressants as a cure.

Think about how some individuals were depressed prior to taking medication and symptom free after they stopped. Consequently, they may testify that antidepressants cured their depression.

On the surface, this may appear logical. However, imagine that you sprained your ankle and took an analgesic for three days, after which your pain was gone. You wouldn't say the medication cured the symptom. You'd say that your ankle healed during that time. This is an example of how confounding variables (say, the body's ability to heal and the time necessary) can properly account for an outcome, rather than thinking that painkillers cured your sprained ankle.

Likewise, if you found out that while taking antidepressants dysthymic persons found a job, finalized a divorce and started a new relationship, addressed a long-standing medical issue, repaired a broken relationship with a distant parent, or changed their way of thinking about a stressful situation, you'd factor in those variables as possible reasons for symptom elimination.

Sometimes, antidepressants suppress symptoms until a stressful situation, or people's thinking, change. Still, people may inaccurately conclude that antidepressants cured their depression when other factors can account for the improvement. Given a particular belief it may be difficult to recognize – or even know to look for – different reasons for symptom appearance and elimination.

When people have a mindset that all medications are curative, they may misperceive antidepressants and arrive at incorrect conclusions.

Some people mistakenly believe antidepressants will make them cheerful. Yet, like an absence of debt doesn't mean that someone is wealthy, antidepressants or mood leveling medications may just reduce the probability of rolling downhill into valleys of despair.

When people don't feel happy while on antidepressants, they may compound their sadness and frustration due to their unrealistic expectations.

Prescribers combat this undesirable possibility by informing their patients that antidepressants are just *anti*-depression, not happiness seeds.

There's another concern related to medically suppressing symptoms.

Take a moment to consider how people drink coffee to help them wake up and get energized for the day. In fact, it works so well that over time even the mere scent of coffee can stir people to action.

Now, consider what might happen when they adjust to caffeine's effects.

People may increase their consumption by bigger and more frequent doses (think short, tall, grande, and venti) or smuggled in by other means, say dark chocolate covered espresso beans.

But something else might happen with an increased intake of caffeine.

Perhaps there's an amplified probability for a physiological and psychological craving and a noticeable withdrawal in its absence.

Analogously, some people may experience particular symptoms when they discontinue antidepressant drug treatment.[80] Thankfully, medical professionals go to great lengths to guard against such circumstances. Regrettably, far too many patients abruptly stop taking their medication and recklessly *not* inform their prescriber. (Predictably, prescribers generally dislike when their patients fiddle with their medication and make decisions they're unqualified to do, such as doubling the dose, not using it as recommended, or taking it with alcohol.)

Employing the pain metaphor once more, consider how you use oven mitts to avoid the hurt that would otherwise occur when you take a hot pan out of the oven. Of course, you'd stop using the oven mitts once you've placed the pan where you desire or after it has cooled.

Yet, if you believe the pan is hot after it has cooled, you'll think, feel, and act in accord with that misperception. So, out of fear that you'll burn your hands, you'll needlessly continue to wear the oven mitts to avoid anticipated sharp and unwelcome pain.

Okay, but there's a catch.

You deny yourself the feedback as to whether you need to keep wearing them. The lack of pain reinforces oven mitt use. If you never test whether you need them, you'll risk wearing them long after it's necessary to do so.

Now, consider how this may apply to medicating dysthymia.

Like oven mitts, antidepressants can reliably reduce or prohibit symptoms. But, consider what may happen when a stressor passes or those in therapy address their ways of thinking associated with symptom development.

People may not know if they'd feel fine if they *didn't* take antidepressants. Some fear what might happen if they stopped (e.g., withdrawal, symptom return), which may prompt continued, but unnecessary, use – especially if dysthymic individuals believe their symptoms have a biochemical or genetic origin. Prescribers properly monitor and protect against such pitfalls.

Individuals often maintain established patterns (e.g., workout regimen, morning coffee, weekend chores). Now, consider how some people take their medication at a particular time, say as part of their morning ritual. Habit-like antidepressant use may minimize the consideration of whether they're needed.

People risk using antidepressants when they no longer need them.

But there's another factor about suppressing symptoms to keep in mind.

Dysthymic persons may equate using antidepressants to taking aspirin or ibuprofen for a recurrent physical concern (e.g., sore back, sciatic nerve pain) – a medical solution to a medical (non-psychological) issue. Some medicate their depression with the belief that it will cure itself like many bodily issues can (e.g., sprained ankle, strained muscles, bloodshot eyes, simple wound). So, when a stressor passes and symptoms go down, some use that as proof that their symptoms self-heal (though they'll reappear). Consequently, many accept the ongoing practice of intermittently medicating their psychological symptoms.

When people misidentify innate factors (e.g., genetic, neurological, hormonal, neurotransmitters) as the cause of their psychologically-based depression, they mistakenly believe they're addressing "the underlying cause" when they medicate their symptoms.

When antidepressants reduce symptoms, those with persistent depressive disorder may counterproductively validate their misconception that an innate cause poisons their mood and antidepressants are the antidote.

While taking medication for temporary issues can be appropriate, it can be problematic when individuals are intermittently, but endlessly, medicating their psychologically-based symptoms – and therefore risk perpetuating them.

Note that it can be immensely challenging for people to be committed to, and vulnerable in, therapy. Like going to the gym, it requires long-term dedication to a difficult, and sometimes painful, workout. (As most learning does.)

After all, patients have to express the thoughts, fears, flaws, imperfect logic, painfully-awkward experiences and other information they've kept classified as top secret, as well as question their beliefs and perceptions of reality.

As if that weren't difficult enough, they have to take ownership for their symptoms and be responsible for implementing solutions. Also, therapy requires a significant investment of time and, for those who are uninsured or underinsured, money. Last, some people face the possible – but needless – embarrassment of seeing a psychologist.

So, similar to how people often sidestep going to the gym, it's unsurprising that people avoid going to therapy.

When people incorrectly believe that innate microscopic factors cause their psychologically-based depression, they can focus on medically quieting their symptoms…and shift their attention away from therapy.

When people medicate their symptoms, they may rationalize a retreat from things they wish to avoid (e.g., changing their thinking or acting), yet believe they're working on addressing the (perceived) cause of their depression.

Consider why this might be more detrimental than it first appears.

To start, think about those who inaccurately believe unchangeable, inborn, organic issues cause their psychologically-based symptoms. Some depressingly perceive themselves as innately-flawed and powerless victims who must endlessly medicate an unstoppably-chronic condition. When antidepressants reduce their symptoms, they appear to confirm their rather dismal suspicions.

Yet, similar to physical symptoms, psychological symptoms can identify that something's wrong, be a call to improve thoughts, actions, and situations, and productively guide assessment and treatment.

Knowing that psychological symptoms are, in part, a feedback mechanism is an empowering and liberating concept, *as it reveals that people can influence their symptoms by changing their thoughts, perspective, and objectives.* When they do, they can build self-confidence and a reassuring sense of peace.

When dysthymic individuals use antidepressants to overpower and silence their symptoms they may stifle their innate feedback mechanism and inadvertently avoid addressing the issues their depression strives to identify. (Perhaps this is similar to putting electrical tape over the annoying warning light on your car's dashboard when you don't have the time, money, or expertise to solve the problem). Consequently, some risk medicating without healing.

Medication may mask manifestations of maladaptive mindsets.

Medical professionals customarily address this potential problem by prescribing antidepressants with the agreement that their patients concurrently work on their psychological concerns in therapy.

Points to Consider

- Antidepressants can humanely and rapidly reduce people's symptoms.
- Medication may anesthetize symptoms that can resurface when the reason for their existence goes unresolved.
- People may mistakenly believe their psychotropic medications cured their depression when it's due to other factors that have changed.
- Individuals may risk using antidepressants longer than necessary.
- When medication alleviates psychologically-based symptoms, people may fortify a misbelief of an innate, organic symptom origin.

- Dysthymic individuals may use antidepressants to suppress symptoms as a way to avoid the challenging process of addressing their psychological symptoms in therapy and changing their way of living.

Healthcare providers prescribe antidepressants that may quickly and effectively subdue painful and restrictive depressive symptoms regardless of their origin. This practice compassionately assists millions of people to do things they may be unlikely to otherwise accomplish.

However, as psychologically-based symptoms can provide profound and helpful clues about people's cognition, perspective, and goals, there are times when merely suppressing them medically can be counterproductive.

Now, consider the challenging process has to happen to broaden people's perspective and treatment of persistent depression.

Modifying Mindsets

When people decide to go out for lunch, they may rapidly ask themselves a number of related questions, "What am I in the mood for?", "What's nearby?", "What has good reviews?", "What's affordable?", "Will it be too busy?", etc. This simple example illustrates an extremely common and necessary practice.

For people to accomplish things, they often have to plan or think ahead about various options and obstacles.

Now, extend this understanding to numerous other – more important and complex – activities, say getting married, buying a house, investing for retirement, going on vacation, or curing a disease.

People theorize in an attempt to cultivate a mindset that allows them to understand, predict, and control countless things, events, situations, and conditions…including depression.

Theories are applied to historical, current, as well as future events. So, for instance, people may recall their frustratingly-unproductive dating history to predict when symptoms will arise and how to avoid or deal with them (e.g., "I shouldn't get overly hopeful about my date tonight because every time I have in the past I got very depressed when it didn't go as expected").

This appears to be a logical and healthy protective exercise.

Yet, when people lack contradictory data, they tend to believe their familiar, pre-existing theory is an accurate perception of reality and follow it.

For example, consider what may happen when individuals believe they're depressed because of insufficient exposure to sunlight, but it's actually because the gloomy weather has been unfavorable for socializing and physical exercise.

They may firmly believe and fortify their mistaken notion and act in accord with it (e.g., ride out their symptoms until spring, buy a full-spectrum light). This practice of holding a misbelief, behaving in agreement with it, and strengthening the belief – as well as its related perils – can occur on a larger scale as well, say within a community or society.

When theories develop too many logical inconsistencies or are unable to illuminate, it may be desirable – indeed mandatory – *to broaden their scope, limit their applicability, or discard them.*

In fact, some of the biggest advances in knowledge are achieved when theories fail under the increasing weight of contradictory evidence. For instance, there was a point in history when people observed the sun dutifully travel from horizon to horizon and they confidently concluded that it revolved around the earth. This belief was a seemingly logical one, alibied by eyewitness testimony as well as earth's undetectable rotation. However, a heliocentric paradigm eventually eclipsed the geocentric one.

Improving mindsets is a perpetual pursuit. It enhances understanding, as well as people's ability to forecast and influence events and circumstances. This is a good and noble endeavor, as education tends to be.

New paradigms may evolve and those theories that have a better fit and are superior in predicting, explaining, and controlling are able to survive and thrive, while the old paradigms wither and die.

Yet, there's a substantial obstacle to overcome.

Just think about how magic tricks and three-dimensional optical illusions playfully deceive you into believing that something impossible is possible. It may be incredibly challenging to determine just how the deception works given your expectation and point of view. However, by taking a different perspective the illusion is shattered, and what was magic and opaque becomes science and transparent. No doubt you've successfully solved a perplexing puzzle or problem by adopting a different angle or using a novel approach.

Just as tricky as it is to figure out an illusion while standing in the same spot physically or cognitively, paradigm shifts (changes in mindset) and the resultant gains in comprehension may be difficult to achieve while looking through an established, but somewhat distorted or partially occluded, theoretical lens.

In addition, when you realize that people's mindset creates their respective perception of reality, you must also note that people can have an *extraordinarily tough time changing their perception of reality*.

After all, doing so seemingly imperils their ability to understand, predict, and control various things – *which can threaten goal achievement, and by extension, survival and thriving*. For instance, merely considering any change in how to work, parent, and invest can bring unnerving levels of indecision and worry.

Changing mindsets can be so unsettling that people dismiss new perspectives with instant intolerance, quickly silencing the conversation as if muting the TV or changing the channel to avoid an abrasive or awkward advertisement.

Now, think about how this relates to depression.

The cause and effect mindset has guided psychological theory, assessment, and treatment from the earliest to the most recent practices. *Fortunately, comprehension and care have dramatically improved over time, and the cause and effect perspective and practices can be usefully applied*.

Yet, you've also come to recognize that some currently-employed cause and effect-based concepts that can be problematic, unsatisfactory, or restricting.

For example,
- While memories may be used to identify the causes of depression, they're often unreliable and prone to misdirection and corruption.
- The notion that insight guarantees change is incorrect.
- The evidence for a chemical imbalance or genetic cause of dysthymia isn't as widespread and potent as people often believe.
- People misperceive the probability of being hardwired for depression.
- The pitfalls of medically suppressing psychological symptoms.
- When people misbelieve their psychologically-based symptoms are genetic or biochemical, they may see themselves as innately

flawed and relatively powerless victims subject to their body's unpredictable whims.

As you've become increasingly aware of the advantages and shortcomings of cause and effect-based assessment and treatment, you can better grasp their accuracy and influence, and not simplify or overextend their applicability.

With that in mind, consider how the cause and effect mindset might obscure and overshadow important facts, and perhaps get you off track by prompting a manhunt for one or more usual suspects believed to be guilty of causing persistent depression.

Perhaps you've detected the clues along the way and have deduced the fugitive factor – the overlooked perspective that can provide a more comprehensive view and treatment of dysthymic disorder.

Sometimes identifying a cause of depression can be a case of mistaken identity. Just think about how the cause and effect models of depression can have a *unidirectional* character. Simply put, something (e.g., genetics, chemical imbalance, improper neuronal structures, malnutrition, interpersonal conflict, stressful childhood, brain injury, job loss, financial difficulties, heartbreak) causes symptoms.

While it's necessary to investigate those relevant factors, there are other things that help decode persistent depression.

Clinical Clues

Dysthymia can be relentlessly painful and interrupt people's ability to function in different areas of life. Within the cause and effect paradigm, the solution should be relatively straightforward. For example, reduce or end depression by finding the cause and then fix, medicate, or cure it, by…

1. Searching for what causes depression (e.g., past events, trauma, unfavorable living conditions, chemical imbalance, genetics, unhealthy thinking, social discord, loneliness, faulty neuronal structures).
2. Address symptom sources; for instance, looking for historical or current triggers, finding symptom patterns, adopting healthier ways of thinking and acting, changing diet, or taking medication.

With that in mind, it's perfectly logical that people would seek treatment without delay and put forth consistent effort to swiftly alleviate their symptoms.

Yet, consider why dysthymic individuals might do the following.

- Frequently fail or forget to take their medication.
- Determinedly hunt for symptom causes, yet never find any.
- Often ignore doing, or bringing in, their therapeutic homework.
- See a series of different therapists, but never make any progress.
- Make late cancellations or repeatedly fail to attend appointments.
- Attend therapy regularly…but remain mostly silent in each session.
- Stop taking antidepressants even though they were reducing symptoms.
- Inaccurately self-diagnose with a different, and perhaps far more serious, condition.

- Speak with unparalleled confidence even when incorrect about their symptoms.
- Continually and determinedly disagree with their therapist's perspective and suggestions.
- Go to weekly therapy appointments for years but have little or no symptom reduction.
- Readily, and sometimes sharply, reject the view that talk therapy can address their symptoms.
- Routinely misdirect therapy by repeatedly starting an archeological dig through ancient memories.
- Neglect to, or outright oppose, changing thoughts or actions that could reduce or eliminate symptoms.
- Terminate therapy before achieving their stated treatment goals, which allows problematic issues and symptoms to persist.
- Rebel against the regular responsibility of taking antidepressants at certain times, yet in doing so sustain their symptoms.
- Forcefully insist that their depression is genetic, neurological, or biochemical in origin when they lack supportive data.
- State that their antidepressants are inadequate or powerless, yet don't seek a change in the type or dosage of their medication.
- Attend treatment frequently, but continually deflect the conversation to irrelevant issues (e.g., the weather, a distant cousin's child).
- Determinedly avoid using antidepressants, even when they explicitly state that they want a relatively rapid decrease in symptoms.
- Schedule consistent, but infrequent, appointments (e.g., once every three months, once a year), and repeat this pattern for many years.
- Report that after trying every type and dosage of antidepressant medication, all have failed to bring about any symptom reduction.

- Cease using medication with a stated fear of potential side effects (even if they haven't experienced any during the entire duration of usage).
- Start therapy at a time of significant symptom intensity. Attend a session or two, then stop treatment, even though (less intense) symptoms remain.
- Avoid getting a comprehensive psychological examination using standardize assessments which can quickly provide data that can be employed to guide and improve treatment.
- State they don't want to become "hooked on drugs" before discussing the possibility of a prescription, when they're prescribed the lowest dose of an antidepressant, take medications only as needed, or they willfully, consistently, and excessively use other drugs, such as caffeine, alcohol, nicotine, or marijuana.

There are times when variation in how individuals respond to medication, depressive lethargy, unskilled therapists, financial limitations, reluctance to be vulnerable, life obligations, or other reasons readily explain much of the above.

The remainder can appear curiously illogical and counterproductive.

Some people perceive and accept these actions as inexplicably mysterious and rare anomalies, that are randomly incompatible with otherwise universally-applicable cause and effect theory and practice. Consequently, they may gloss over such inconsistencies without further investigation.

Yet, there are clinical clues that hint at a missing piece of the puzzle.

If treatment were as straightforward as identifying causes and employing therapeutic interventions, then the preceding would be far less likely to happen. (However, I assure you, these and other similar actions occur quite frequently.)

Here's where another hint may be advantageous.

How might these attention-getting cognitive and behavioral infidelities to the cause and effect mindset be similar to symptoms?

They can act as feedback that identifies problems and prompts improvement.

Operating only within a cause and effect mindset restricts people to a single direction of symptom development: *Something causes depression.*

A concept that can be logical, familiar, and helpful.

Yet, consider if countertherapeutic thoughts and actions aren't arbitrary, inexplicable anomalies, but the lipstick on the collar, the long strands of hair where they shouldn't be, that reveal hazards and limitations which lead people to be less wedded to the cause and effect mindset of symptoms.

If so, then it's easier to revise the understanding that depression can only live within a causal framework.

Intriguing.

Thankfully, there's another perspective of psychological symptoms that can address the shortcomings of the cause and effect model, enable you to see a bigger, clearer picture of dysthymia, explain inconsistencies, as well as broaden understanding and care.

Growth Model

Consider how people may misapply the cause and effect perspective. For instance, some mistakenly believe that their genes cause their obesity, when it's actually their problematic perception, rules, and objectives for eating that need to change. Or, think about those frustrated with their low energy or strength, and wrongly accuse their genetics or biochemistry while liberatingly and narcotically overlook their long-term lack of exercise.

Yet, to assist individuals to become healthier, nutritionists don't dwell on why their clients are overweight or how they became that way. They help people reach their goals by instilling new conceptions about food, more appropriate eating guidelines, awareness of how their body works, and rewarding nutritious eating as well as other Olympian-like behaviors.

Likewise, personal trainers aren't preoccupied by how their clients weakened over time. Rather, they provide data and observations that help people develop safer and more effective workout rules and habits.

Nutritionists and personal trainers emphasize and increase people's strengths, instill healthier perspectives, thoughts, actions, and goals, encourage individual choice and responsibility, as well as enthusiastically gold star their clients' successes to help them maintain a better way of living.

With that in mind, consider a perspective that may address some of the shortcomings of the cause and effect mindset: The Growth Model.

The growth model avoids the "mental illness" label for psychologically-distressed conditions and instead views symptoms as a function of development; for instance, improper or insufficient learning that can be remedied by studying the thoughts and behavior of non-symptomatic individuals.[81]

This can be a more positive conception of people and symptoms.

Instead of searching for and lingering over negative, unchangeable past events or conditions, or endlessly re-examining present and past symptoms, the growth model enables and advocates a more optimistic approach.

Practitioners employing the growth model focus and build on people's strengths and encourage healthier thoughts, behaviors, and objectives.[82]

Growth model practitioners accentuate a positive future; for example, how people can think and act differently to avoid symptom development. They emphasize learning and help individuals acquire more appropriate perspectives, thoughts, behaviors, social connections, expectations, and healthier goals, for instance, that enable them to reduce their symptoms and reach their potential.

Rather than dwell on people's symptoms and defining them as victims, growth model practitioners tend to view them as discouraged (e.g., after failing due to inadequate rules), and in need of developing a more positive attitude and foster their awareness of their ability to influence outcomes.[83]

Given the emphasis on learning, growth model practitioners refrain from fully relying on psychotropic medications.[84]

Medication can accelerate therapeutic progress by relieving distracting and demotivating symptoms. However, as they can provide feedback, sedating them has risks; such as giving a false impression of a cure, or decreasing the ability to detect symptom-related thoughts, actions, situations, and desires.

But there's something else you should know.

Consider how firefighters must quickly determine a fire's location and how to put it out safely. Likewise, the cause and effect perspective prompts people to rapidly identify symptom causes and extinguish symptoms. This can be a wonderful and healthy thing. While the growth perspective can be employed similarly, also think about the benefits of the following.

The growth model can authorize and advocate therapy for personal improvement as well as for preventive measures.[85]

In this way, the growth model is somewhat akin to fire prevention skills – protectively avoiding symptomatic fires, rather than waiting for them to ignite and then having to put them out. While insurance companies helpfully subsidize therapy after symptoms arise, funding preventive pursuits is less common.

The growth model provides an additional, complementary voice to the thunderous chorus of the cause and effect mindset.

The growth model allows for the understanding of how people are active, creative, social, make choices, and move toward a range of objectives.[86]

Knowing that, can you see how it can avoid many of the difficulties and uncertainties associated with the causal mindset of symptoms?

Well, for example, the cause and effect models view internal and external factors as triggers of depression (and rightly so in many cases)...*but may not examine and explore the telling extent symptoms influence others.*

Hmm.

What follows is a meticulous investigation of a mazelike mystery. With the growth model in mind, you're better able to detect clues and contemplate evidence to decipher how and why dysthymia can rise, survive, and thrive.

Often, this requires adapting or discarding mistaken beliefs that may misdirect or limit understanding and treatment.

For instance, as the growth model recognizes humans as curious, active decision makers who strive for various outcomes, there's a popular cause and effect principle and practice you have to review and revise.

Perceiving People as Passive

The influential cause and effect mindset is as familiar, socially accepted, and encouraged as well-advertised cola. (Odds are, you've seen the concept in countless movies and TV shows in which something happened to the protagonist who then acts differently because of it.)

So, it's unsurprising that people often identify internal or external factors as triggering dysthymia. Accordingly, much of talk therapy as well as medical interventions use procedures and prescriptions to relieve symptoms by searching for and addressing the perceived cause.

Yet, along the way various clues have exposed some boundaries of the cause and effect mindset…and hinted at something else you should consider.

When depression is deemed to be purely the result of external causes, such as job loss, a broken heart, or rejection by friends or family – or from internal matters, say genetic abnormalities, chemical imbalances, or "bad wiring" –dysthymic persons may be perceived as passive victims.

Knowing that, consider how the following risks might arise.

- Individuals remain unaware that they can influence their symptoms.
- Needless pruning of hardy branches of assessment and treatment that could lessen symptoms.
- People's misery mounts when they maintain a self-image of being unfortunate and powerless casualties of their depression.
- Others' expectations are lowered, perhaps to the point of acceptance or pity, which hinders or halts the healing process, or heightens symptoms.
- Decreased probability people will take responsibility for improving their moods and mindset (e.g., "I'm sure it's genetic [or due to a chemical imbalance]; so how can I possibly change it [or why try]?").

- Obscuring the depressed person's role in symptom development, for instance, the words, attitude, and actions that led to being fired from work, spurned by a lover, or ridiculed and rejected by friends.

Perceiving depressed individuals only as passive victims has striking limitations and countertherapeutic risks.

So, think about what may allow greater comprehension and care.

To start, consider what you might do when you see a sad family member, friend, or a co-worker crying. Perhaps you'd stop what you're doing, listen attentively, ask what's wrong, reassure, and do your best to bring cheer through an act of kindness, for example, offering encouragement, buying a small gift, or helping out with some chore.

Such responses offer you a peak around an obscuring, opaque curtain of the cause and effect mindset and uncover a crucial clue.

Depressive symptoms can influence others and provoke particular outcomes in the social environment.

With that in mind, think about what fundamental human characteristic suddenly becomes a valuable informant (and possibly a prime suspect).

Goal Orientation

To comprehend persistent depressive disorder, you must have a basic understanding of how and why people act. But as life doesn't come with an instruction manual, this can be a bit of a bind. Fortunately, theorizing, trial and error, instruction, observation, experimentation, and searching for existing facts and acts enable insight and knowledge.

Here's an interesting illustration: Did you know that when accumulated scientific data and research has failed to provide a solution, pharmaceutical companies have sent ethnobotanists and physicians to remote areas around the globe to talk to "medicine men" in small tribes, clans, or other groups to find what procedures or materials (such as vegetation or animal products) they use to successfully treat various afflictions? The collected information provides a starting point for developing pharmaceuticals and medical treatments.[87]

Like starlets in Hollywood, helpful clues and solutions are abundant and patiently waiting to be discovered.

For example, just think about how for life to exist Mother Nature used a brute force attack of infinite trials to figure out innumerable solutions. The formulas that worked, remained.[88] Consequently, nature can be a bountiful reservoir from which to draw answers and develop insight that can be productively employed (biomimicry).[89]

While some of nature's hints and answers are quite usable, others are indirectly or partly so (e.g., while airplanes don't flap their wings, they use some avian aerodynamic principles), and the remainder are inapplicable.

Sometimes the biggest clues exist in the smallest things. Consider how deoxyribonucleic acid (DNA) is pregnant with genetic information about how to construct and operate living things. And, perhaps the most primary component of DNA is the gene, as it's the basic element of transferring data through generations and the smallest unit of natural selection.[90]

Now, consider what clues DNA may provide about persistent depression.

Fascinatingly, DNA contains legions of genes (broadly, sets of instructions) that when followed produce a state, function, or construction. Notably, genes are often in silent life or death territorial disputes – the victors remain, the vanquished don't. (Or, if you prefer, you can think of it like genetic musical chairs in which some genes loose out to dominant or more numerous others.)

Now, consider how a plant's genes contain information for it to achieve a particular height, width, root depth, color, scent, leaf configuration, capacity for replication, and function. When there's enough sunlight, nutrients, irrigation, and few or no predators, the plant may reach its genetically-engraved ideal, although that may not be perfect by other measures, say what's needed to sufficiently address environmental threats.

But if the plant is in the shade when it needs sunlight, suffers through a harsh rainy season or two, is in inhospitable soil, or insects or disease infest it, the plant never quite reaches the goal written in its genetic recipe. It may be a bit shorter, rougher around the edges, unable to reproduce, or influenced in other ways. In fact, there are countless examples of how physical and social environments impact life.

For instance:
- Plants develop protective thorns.[91] Also, they may gently genuflect in the direction of sunlight (positive phototropism).[92]
- Animals grow a protective carpet of thick fur,[93] gradually transform colors or patterns and blend in better with the environment,[94] or move in groups thus better able to protect each member against attack.[95]
- Bacteria evolve and become resistant to antibiotics.[96]

Of course, these adaptations reflect the process of mutation and natural selection rather than conscious choice. Even so, consider what may be applied to understanding humanity – and perhaps reveal vital elements of persistent depressive disorder.

First, genes provide instructions for various aims of form and function. If certain targets are not met, then plants and animals do not grow, survive, or reproduce.

In order for life to exist, there must be goals. For this reason, the concept of goal orientation is crucial.

Second, surroundings strongly impact the ability to survive, way of living, visible characteristics, and reproduction.

Social and physical environments are incredibly influential.

Third, while goals can be ideals (e.g., a plant forms and operates in a specific way to move water, aka transpiration), and some DNA-based objectives need to be met completely, in some cases achieving a genetically-imprinted ideal is not needed, or more likely, probable.

For instance, if a plant's leaf or an animal's height is mildly different from the DNA-based target, either may be able to live and thrive. After all, there might be very little life if plants and animals had to entirely meet all of their genetically-based objectives. (Remarkably, small changes over many years can make massive differences.[97] Just consider the giraffe's neck as proof.[98])

There can be a window of acceptable development, and times when approximate goal achievement is satisfactory. (In fact, the phrase "survival of the good enough" is more accurate than "survival of the fittest".[99])

Fourth, when DNA-based instructions are unable to create form, physiology, or guide behavior to suitably deal with physical and social threats, requirements, stressors, and changes, those life forms in which mutation edits their genome may be better able to adapt, thus survive and thrive.[100]

It may be essential to deviate from original, or ideal, goals.

Simply put, DNA can reveal the need and supreme importance of goals.

Additional illustrations may further ease the shift in focus from basic biology and genetics to human presentation and psychology.

In the preceding examples, you saw that DNA contains goals for how plants and animals function and present. Likewise, people have DNA-based physical and physiological goals, such as height, respiration, and heart rate. Or, consider how beginning at conception, the DNA-based objectives instruct how and where bodily organs grow and how they work.

You also read how physical environments can impact plant and animal presentation. The same is true for people. For example, Tibetan Sherpas who live high in the mountains have larger lung capacity.[101] Consequently, they're better able to breathe the rarefied air, presumably because those who had lesser capacity lungs did not survive to reproduce. Other long-term changes include skin color[102] and hair thickness.[103] Physical environments can influence function and visible characteristics in the long-term. In fact, humans are still evolving.[104]

Goals are important in human functioning. They exist even at the microscopic level. But this doesn't mean that people are powerless against their DNA and all their thoughts, actions, and emotions are genetic in origin.

Think about the importance of individual choice in goal selection and achievement. In fact, people's selected objectives can be far more persuasive – and revealing – than you may first suspect.

For example, consider what it means when people identify themselves by their professions (e.g., engineer, construction worker), possessions (e.g., a Harley-Davidson or pit bull owner), hobbies (e.g., yachtsman, New York Yankees fan), personality characteristics (e.g., aggressive, persistent), or physical build (e.g., muscular, athletic).

People define themselves by their goals and the means to achieve them.

Individuals' objectives are omnipresent, inescapable, and complex.

For instance, to become a lawyer is an aspiration, but it can also be a means to earn an income to achieve other ambitions such as home ownership. Some people want to have a certain physique, which then may be used to get attention that leads to a date and, eventually, marriage.

One goal can assist in achieving other goals.

Next, consider what's demonstrated when individuals are married and employed, while raising a family and pursuing a hobby.

People may simultaneously move toward multiple goals.

Now, reflect on how you progressed upward through grade levels during your schooling, and what that reports.

People may pursue and achieve goals serially.

There are revealing clues everywhere about people's goals. For instance, consider what sports, symphonies, and sex reveal.

As individuals can cooperatively strive to attain the same objective, *goals can be collective*.

Plants and animals may be able to survive and thrive when they fall within a window of developing toward their genetically-engraved goals. With that in mind, think about how some youth idealistically intend, "When I grow up, I'm going to live in a brick house with five bedrooms, a swimming pool, and a six-car garage", but eventually are perfectly happy to live in a vinyl sided three-bedroom house with a two-car garage (and, perhaps, use their neighbor's pool). Or, thinking about this a bit differently, you may not always have your favorite meal, but you can satisfy your hunger from what you're able to eat (e.g., you want an elaborate, healthy meal, but only a drive-through restaurant is open).

There are times when approximate goal achievement can be good enough.

Positive genetic mutations can increase a plant or animal's ability to survive and thrive as it can slightly change a DNA-based objective that allows an advantage, a beneficial adaptation to the environment (e.g., bigger leaves or greater height that allow better access to sunlight).

Knowing this, recognize that dreams to be a professional football player, doctor, movie star, business owner, or parent, are often interrupted or turn into nightmarish situations for various reasons. Even ordinary daily life can be stormed by some natural disaster such as when a hurricane, flood, or fire destroys people's homes and possessions. However, people may accept the fact they must deviate from their original objectives, select new ones, and adapt to their environmental stressors.

Goal revision may be necessary and healthy.

Now, consider what the following suggests.

When plants and animals exist in a threatening social environment (e.g., predators, infestation) it can influence their presentation and actions. Likewise, when they live in a healthy and hospitable environment they may respond in particular ways (e.g., plants that move toward sunlight, dogs that excitedly greet their caring owners [cats, not so much]).

With that in mind, recognize how individuals don't present and act the same when they're in the following situations.
- A creepy, dark alley.
- An amusement park.
- Their cozy living rooms.
- A business meeting.
- A rowdy sporting event.
- A solemn courtroom.
- An insufferable relative's house.

People act differently in each location. For instance, they may be far more vigilant in a dark alley to reach their goal of physical and financial safety, filter their speech at a business meeting to remain employed, or sit far away from a know-it-all relative to avoid conflict, a headache, or a painfully long argument with their spouse during the ride home.

Reflect on how people can have a sudden change in discourse and disposition when a police officer, young child, nosy neighbor, jealous spouse, or clergy member approaches. Indeed, it's likely that you've said things to friends that you'd never say to your family (and vice versa).

Individuals can adapt to situational requirements and act in accord with their goals.

Next, consider how fashion, fads, cultural rules, political zeitgeist, and religious customs are common large-scale social influences on an individual's thought, emotion, action, presentation, and goal selection.

Relatedly, consider what the following examples illustrate.
- A supervisor who barks, "Finish the report!"
- An army colonel ordering soldiers to "Go over the hill and storm the enemy's location!"

- Doctors telling patients to take their medication every four hours.
- Individuals on the verge of orgasm breathlessly directing their partner to "keep going".

People can have goals thrust upon them.

Just think about how familial coaxing and presumptions in sports, academics, or employment affect income, health, status, and the ability to marry and have children. Or, reflect on how the ownership of a family business might guide children to "carry on the tradition."

Familial factors such as parental encouragement or discouragement can influence goal selection.[105] Again, you can see how social environments influence people's behavior, appearance, and even the ability to live.

Social environments are stealthily and powerfully persuasive.[106]

Next, think about what scratching an itch, getting to the movie theater on time, working at a particular company, having sex, becoming an accountant, and saving a certain amount for retirement reveal.

Goals can be any size or type.

It's monumentally important to recognize that people are perpetually in the process of setting and achieving objectives on all levels of functioning, in various areas of life, serially and simultaneously, individually and collectively.

Goals have enormous importance in human functioning.

(If you were wondering why I went into this so much, it's because this fact is supremely relevant in comprehending and treating dysthymic depression. It also reinforces why perceiving people as passive can be problematic.)

Life Tasks

Many animal species are born with a high percentage of their brains complete with instincts and preprogrammed behavior, which is in contrast to humans who are born with a lower percentage of their brains developed.[107] Some of those innately-inscribed behaviors animals demonstrate include swimming, walking, building nests, spinning webs, constructing dams, courting, and migration, all of which increase the likelihood of survival and reproduction – especially for those species whose newborns that have to fend for and feed themselves.

For people, some goals and the means to achieve them are set in their DNA and don't require conscious thought (e.g., blink reflex to protect the eyes). However, other objectives and the ways to attain them aren't programmed (e.g., what job to get and how to do it). *There are areas of life that have no inborn solution or behavior.* Consequently, people are faced with the formidable challenge of what goals to select and how to achieve them.

The inescapable arenas of functioning which people are obligated to address, but which lack innate solutions, are called the Life Tasks.

In infancy and young adulthood, parents and other adults may provide food, clothing, shelter, and safety. But, later in life these physical and physiological needs are often met through finding some means of generating an income so that people can buy the possessions that satisfy these requirements as well as make life more bearable and enjoyable.

This challenge is broadly defined as the *Work Task*.[108]

It must be answered in some capacity, usually through having an objective (e.g., be a surgeon) and a means to achieve it (medical school).

However, the work task doesn't require employment.

After all, some people survive from what they can beg, borrow, or steal, others might win the lottery or receive an enormous inheritance, or some choose to live off of others. (Indeed, there are many adults who live with their parents well after the healthy expiration date.)

Also, keep in mind that the work task may be addressed individually or collectively (e.g., both spouses work to pay the mortgage and put food on the table). Note as well that the work task also represents what people do when they're not working, say being on vacation or enjoying hobbies.

The *Social Task* requires individuals to address how they wish to connect and interact with others, if at all.[109] Given what you know about the immense importance of goal orientation and achievement, consider how social interaction, interdependence, communication, guidance, culture, technology, and collective action can be related to this task.

The second part of the social task is how individuals address the need to belong. For instance, people may experience a sense of belonging with their families, friends, co-workers, teammates, or club members.

Life also presents the *Sexual Task*, which consists of four subtasks.[110]

Sexual role definition: This is how people define what it is to be a man or a woman. It's a social role as well as a biological one.

Sex role identification: This is how individuals see themselves in regard to what is a man or a woman, or the ideal man or woman. Individual perspective, as well as what others, and culture in general define as masculine and feminine shape this perception.

Sexual development: This is how individuals achieve sexual development milestones and how they react to these events.

Sexual behavior: This is everything related to sex: sexual feelings, appropriate and inappropriate sexual behavior, and includes selecting and achieving goals in such areas as dating, courtship, marriage, family, sexual behavior, and choice of sexual partner, which represent physical closeness as well as mutual and compatible goals, emotional connection, teamwork, acceptance, and love.

Similar to the other tasks, the sexual task doesn't have to be answered in the affirmative, as people may select goals of isolation and alienation.

The *Self Task* is how people define and evaluate themselves as if from a third person perspective, such as "I _____ me."[111] This task has several subtasks:

Survival: Be it biological, psychological, or social. In other words, it's people's physical and mental health, self-esteem, and sense of belonging.

Body image: This is how people perceive their physical beings and whether they like or dislike their bodies.

Opinion: This is what people think of themselves, such as "I like myself." They can accept and admire part of themselves, for example, their work self, while hating another part, say their social self.

Self-evaluation: This represents the self-conception of being received with warmth and compassion, discouragement, or indifference.

The *Spiritual Task* is how people give meaning to their lives such as, what they do about such things as religion, where they see their place in the universe, and what they think about life and death.[112] Note that this task does not require the assumption of invisible supernatural beings, magical thinking, a belief in mythology or the mystical.

The *Parenting and Family Task* is how people perceive their families and interact (or not) with them.[113]

The Work Task, Social Task, Sexual Task, Self Task, Spiritual Task, and the Parenting and Family Task comprise the Life Tasks, which are the areas of functioning that do not have inborn answers and people must subjectively address by selecting goals and the means to achieve them.

Perhaps you're thinking that unemployment or an unsatisfying career would create symptoms. Or, you may be certain that loneliness and a lack of sexual intimacy can be depressing to the point of developing dysthymia. Possibly you've decided that those who do not accept themselves must be sad. Maybe you conclude that those who lack spiritual connection can feel bored, alienated, empty, and lethargic. Last, you may decide that rejecting parents, distant siblings, disengaged children, or being childless could produce persistent depressive disorder.

The above appear to be straightforward, logical reasons for depression. Yet, they fall within the cause and effect perspective of symptom development.

Fair enough. But consider how the life tasks can relate to depression from the growth model perspective. Keep in mind that people are goal oriented and must address the inescapable life tasks; however, what think about what position they're born into that can make the process so intimidating and prone to failure.

Inferiority

People are goal-oriented beings. But consider their top two objectives.

To survive and thrive. (This may sound familiar.)

However, there's a problem. While people have DNA-based goals of form and function, they don't have sufficient innate solutions to fully address the life tasks. They must learn how to attain their needs and desires. Yet, consider what unavoidable initial condition can make that pursuit exponentially more difficult.

Everyone starts life in an inferior position.

Children are born unknowing of what to do to survive and thrive. They lack significant communication skills, are relatively weak, etc. So, they must rely on others to achieve vital objectives, such as being clothed, fed, and sheltered.

Also, consider how kids often compare their intelligence, physical and social skills, and judgment, to those around them. Usually, the comparison is made with those who are older, such as parents, teachers, and siblings.

Now, think about how this can be problematic.

From children's point of view, nearly everyone else around them is stronger, smarter, taller, better able to communicate, more skillful at predicting and controlling the environment, and make wiser decisions, therefore more capable at setting and attaining countless objectives.

Knowing this, consider what may happen when children assess their skills and knowledge in comparison to others – or fail to achieve their goals.

While it's true that infants are oblivious of their position, as children develop they become increasingly aware of their incompetence and reliance on others, *which may breed a dreadful sense of inadequacy.*[114] Simply, they may experience distressing inferiority feelings, not only in comparison to others, but also to the task at hand.

For example, imagine a child sitting on a training wheel-equipped bicycle watching the older neighborhood kids doing impressive and desired feats, such as rapidly making turns by leaning their bikes at severe angles. The young child feels comparatively inferior and concerned, as incompetence not only prohibits riding the bicycle more effectively and enjoyably, but also jeopardizes things of incredible value: *self-esteem, status, acceptance, and way of living*. If the child keeps riding the training-wheeled bicycle, inferiority feelings may continue or intensify.

Yet, those nagging feelings don't inescapably lead to unending frustration or sadness. The child can be driven to eliminate those unpleasant sensations.

Okay, but think about how the kid can do that.

Perhaps choosing to not make the comparison, or giving up bicycle riding, or *advantageously using those distressing inferiority feelings to overcome them*.

Indeed, when children want to attempt an activity – and perhaps wish to master it – they can be thoroughly relentless in their efforts. Parents are usually familiar with their kids' animated and ceaseless pleas, "Let me do it! I want to try! Please! PLEASE! *PUULLLEASSE!!*".

Accordingly, in an effort to be liberated from the objectionable inferiority feelings, the child may prod the parents to remove the stabilizing wheels. Note that intense emotions can fuel sustained attempts through episodes of failure, hurt, and colorful bruising, to learn to ride the bicycle without training wheels.

So, despite a number of painful falls and scrapes, the child's fiery enthusiasm to succeed at an intimidating task and eliminate stressful feelings of inadequacy, combined with parental support and guidance, virtually guarantee success.

Terrific. More important, consider how the kid may perceive this outcome.

After learning to ride the bicycle, the child no longer feels inferior to the task. This elimination of unpleasant sensations, coupled with goal-attainment that reduces the risk to self-esteem, status, acceptance, and way of living, is as pleasurably-elevating and reinforcing as winning a gold medal.

Fair enough. Now consider how this example is relevant to depressed adults.

Think about how as children grow into adulthood they're responsible for learning how to select and achieve countless goals in a never-ending onslaught of tasks. Anything from the small (e.g., learning how to use a new phone) to inescapable obligations (e.g., schooling, parenting, paying bills) to social interactions (e.g., friendships, sexual relationships), and so on.

Given that people have to set and achieve so many goals in life, but are born with an insufficient innate knowledge of the world and how to act within it, they can be motivated to compensate for this shortcoming.[115]

Yet, acquiring assorted abilities isn't quick and easy. People take many years to learn life sustaining skills (e.g., language, socializing, trade apprenticeship).

Now, imagine those who want to support a family, and go to medical school to fulfill that objective. However, they've taken on such overwhelming student loan debt that they paradoxically imperil their ability to have a family.

Ironically, some goal-directed pursuits to overcome inferiority can jeopardize themselves. (Think about what this hints about depression.)

In addition, note how people may feel inadequate when they're denied a better position at work, unsuccessful at developing new friendships, unable to become sexually aroused or achieve orgasm, or experience disastrous intimate relationships. Failure to attain goals and insufferable inferiority feelings are profoundly and, at times, treacherously intertwined.

People may experience inferiority feelings associated with challenges in various areas of functioning and throughout the life span.

Now consider when inferiority feelings aren't a bad thing.

First, consider how life poses various tasks and opportunities. Some are dangerous or deadly. Consequently, individuals must assess the lethality of the situation as well as gauge their ability to reach the goal.

Overconfident people may take action that's beyond their ability and soon get into trouble. After all, those who have self-confidence in a task they've never accomplished might be grimly mistaken. (YouTube has an abundance of such instantly-regrettable moments.)

Yet, when people accurately assess that they lack the skill to reach a goal, they can peaceably accept what they cannot do and not pursue it (e.g., "I'm not capable of doing electrical work [swim that far underwater, go rock climbing]").

So, consider what this says about inferiority feelings.
1. They aren't necessarily detrimental.
2. They can provide insightful understanding that beneficially guides behavior, say by preventing people from doing dangerous things.
3. They can be a means to assist survival and thriving.
4. While inferiority feelings can be inversely related to self-confidence, they do not have to be associated with symptom production.

Inferiority feelings that stop people's movement may aid goal achievement.

Second, as the earlier example of the bicycle-riding child illustrated, intense inferiority feelings can sustain prolonged and determined attempts. *People can use inferiority feelings to motivate themselves to acquire goal-achieving skills.*

Individuals can determinedly try to attain a desired objective – whether or not they actually do. For instance, they may endlessly try to lose weight, master their golf game, attempt to run a marathon, or learn to play the piano.

Inferiority feelings can generate tension that can be used to prompt immobility or movement.[116]

So, what's the deciding factor as to whether inferiority feelings stop people from action or fuel the fire of determination? (Hint: Goal orientation.)

It's not inferiority feelings that are detrimental to functioning, but how individuals choose to handle them.[117] In other words, it's people's preferences (i.e., goals, rules) that influence what they do. This is a hugely important fact.

Now consider how inferiority feelings are associated with dysthymia.

People may judge themselves as inferior to others (e.g., occupationally, physically, financially, sexually), and conclude they've failed to reach suitable goals – sadly behind and unable to catch up, let alone ever be in the lead.

When people believe they cannot grasp their desires, they can be flooded by a sinking feeling of being hopeless, helpless, and powerless, swamped by life's challenges, and fearful they've irretrievably torpedoed self-esteem, status, acceptance, and their way of living. Dysthymic persons may multiply their inferiority feelings when they believe they're unable to predict, control, or comprehend their symptoms.

But it doesn't stop there. They can counterproductively psych themselves out of healthier action and unintentionally extend their symptoms.

Inferiority feelings and depression can make the barrier to successful action appear insurmountably high. This can paradoxically prohibit people from goal-achieving action that's genuinely within their ability and would free them from the imprisoning self-perception of inadequacy.

Often, people say they *would* approach some task or challenge *if only* they had confidence that they could be successful at it.

This creates a paralyzing paradox. Now, consider the bicycle-riding child example: When the kid eventually learned to ride the bike without training wheels, there's proof of ability in reaching a personal – as well as socially acceptable and encouraged – goal. This achievement builds and fortifies confidence in bicycle riding. This leads you to a telling conclusion.

Confidence can be achieved only after reaching a goal and proving proficiency to oneself – until then there's just a desire and the drive to attain it.

When people make confidence the only key that unlocks the door for action, they effectively *prohibit* the effort necessary to prove their ability and foster self-confidence. This "cart before the horse" thinking is troublesome.

Those who want confidence before proving their goal-achieving ability to themselves create a confining contradiction.

But, there's something else to keep in mind.

It doesn't matter if everyone else in the world believes in someone's abilities, *the person must believe it*. Therefore, even if depressed individuals prove their capability to others, *until they're able to recognize their abilities and successes, they can maintain counterproductive inferiority feelings*.

Relatedly, there's a highly-desirable state when people are able to perceive their goal-achieving proficiency.

Competence

Odds are, you successfully did more than one of the following.
- Plant a garden.
- Run a marathon.
- Conquer a video game.
- Get your house in order.
- Ask for a raise and get it.
- Host a party that went well.
- Raise children appropriately.
- Learn how to play a new song.
- Finish a challenging or lengthy book.
- Pass an exam you thought you failed.
- Successfully ask someone you desired for a date.
- Complete the Sunday *New York Times* crossword puzzle.

Now, here's a more informative question: How did you feel after doing it?

While people can have the drive to achieve their objectives, they aren't emotionless robots who mechanically complete tasks. There's much more to it – and vibrant clues hide like Easter eggs for you to discover.

When people fail to fulfill their desires, they may feel painfully inadequate, frustrated, angry, or depressed. Obversely, when they reach their goals they prove to themselves that they're capable of achieving their objectives and *experience a reinforcing, favorable, and fortifying sense of competency* (e.g., feel relieved, proud, peaceful, happy, energized, liberated).

Next, consider why people feel enjoyable emotional and sensory experiences after reaching their objectives.

Hint #1: Humanity's two biggest goals: To survive and thrive.

Hint #2: Eating and having sex are required for survival and thriving.

Hint #3: Eating and sex are two of the most innately-reinforced behaviors.

The body's capacity to experience such intensely-pleasurable feelings while eating and having sex (as well as disagreeable ones for not) provides additional evidence of the significance of goal achievement in the ability to survive and thrive. This, perhaps on a lesser scale, can apply to other activities.

The affirming and pleasurable sensation of competence reinforces – and testifies to the importance of – setting and achieving goals.

This gratifying feeling can be from attaining tiny aims (e.g., squashing that pesky mosquito in your bedroom when you're trying to sleep) to life changing ambitions (e.g., getting married, having a child, securing your dream job).

There are a range of sensations, depending on how individuals perceive what they've accomplished. For example, some feel terrific after losing weight or graduating college, whereas others who achieve the same aren't so overjoyed.

A sense of competency occurs legitimately *only* when people can recognize and accept accountability for attaining a goal (rather than viewing it as due to luck or someone else's kindness or effort – as dysthymic persons tend to do).

Next, consider what the following suggests about competency.

Proving to others the ability to do a task well may enable employment, exhibiting good parenting skills might put one's apprehensive spouse at ease before leaving the house, or knowing how to beautifully play a musical instrument or sing a song can enthrall, exhilarate, and enrapture an audience.

Demonstrating competency can have a social factor.

(Indeed, sometimes people can be like a kid who just mastered a card trick and wants to show – and amaze – everyone else with that attention-getting, acceptance-ensuring, and status-improving, proficiency. People may show off their skills, indirect signs of competency [e.g. an expensive Mercedes, summer cottage], awards from work, sports trophies, or trophy spouse, for example.)

To survive and thrive, people must be competent in many and diverse ways.

Like celebrities and athletes with multiple nicknames, or the Robert, Rob, Robby, Bob, Bobby variants of the same name, competence is also known as "significance" (these terms – like others in this book – are used interchangeably to avoid annoying repetition.) Making this a bit more challenging, like love and beauty, there's more than one definition.

First, the placement and movement along the significance yardstick can be defined *objectively*. This can be something that has a standardized measure and is universal, for example, being able to run a marathon, becoming a parent, or saving a certain amount for retirement.

Second, the feeling of competence can be a *subjective* personal preference that others may not hold or value (e.g., being religious, volunteering regularly, the ability to tolerate alcohol or illegal drugs).

Third, people can assess their level of competence in *comparison* with others in innumerable ways, both subjectively and objectively (e.g., having many sexual partners or few, being the center of attention at a party or mingling with stealthy caution, having a better house or car than the neighbors). *There's often a social-comparison characteristic in how people define significance.*

Given the countless variables involved (e.g., goals, subgoals, strategies) and how they can be measured objectively and subjectively, *each person may have a uniquely-defined constellation of competency.*

Think of this like athletes who hold the same desire of being on a team, but each strives to be competent in a specific position and employs different tactics.

To survive and thrive, people must face the life tasks and seek significance in personally-defined ways they value and believe will help them, which creates countless self-portraits of competence.

For instance:
- Safety, e.g., "I live in a nice neighborhood, and can afford healthy practices for my family, and I exercise regularly."

- Satiation, e.g., "I believe that I know enough about my job (have enjoyed my life, have a sufficient amount of sex with my spouse, etc.)"
- Being, or possessing things that are, "good enough", e.g., "I live in a suitable home (without having to live in a mansion)", "I have enough in my retirement savings."
- Superiority compared to others, e.g., "I have a better job (nicer kids, larger home, more money, healthier) than my siblings."
- Proficiency, e.g., "I know that I'm good at playing tennis (keeping my home clean and running well, being a parent, etc.)."

Like appetite, the desire for significance extends throughout the life span and varies in meaning given its personal description. For example, your friends, siblings, or neighbors define competence differently than you.

The subjective, idiosyncratic aspect of competency is exceptionally relevant in understanding and treating persistent depression.

To start, what might you conclude from students who are happy performing well, but don't wish to be the class valedictorian?

At first, you might think this discredits the concept of significance. But, imagine you found out that they view their position as ideal in a different way, say having the best balance among social life, relaxation time, hobbies, and studying (and avoid being the center of attention if they were the valedictorian).

People don't need to be the best to consider themselves competent.

But there's something else to note.

Envision people in the shallow end of a pool who feel inadequate and push themselves after seeing others swimming in the deep end, or prompt others to assist them to become competent in swimming in deeper water. Once they learn how, they feel a sense of competency, as they proved to themselves their ability to achieve their goal. This is an example of direct compensation.[118]

Okay, but people don't always face challenges head on.

Just think about when kids are poor at athletics, for instance. They may work hard and achieve excellence in a different realm of functioning, such as academics. Or, those who don't do well academically or athletically may be "too cool" to study, mock their teachers, and use their time in school to socialize and be the most popular. This is an illustration of indirect compensation.[119]

At this point, you have a fair number of clues to figure out why personally-defined pursuit of competence is important to know.

Dysthymic persons strive toward goals they believe grant significance.[120]

Okay, but think about the catch.

When people lack a sense of competency they may feel too discouraged to select and move toward everyday goals, fearing failure and its bitter social penalties. To them, life's challenges appear like that individual they're attract to, yet who they deem as agonizingly out of their league. Depressive lethargy, irritation, and baritone self-criticism may further demotivate and immobilize.

This doesn't mean dysthymic persons don't pursue competence. But given its subjective quality, think about how some individuals cultivate a personal description of it so they can be significant in a way they define.

The relationship between an individual and the social group is complex – particularly so for neurotic depression. Next, consider what human necessity can reveal the social aspect and impact on a person's existence.

Lifelong Socialization

Consider when a person becomes independent from others.

Indisputably, it cannot occur prior to birth given how, for example, a mother's health, diet, smoking, drug and alcohol use influence fetal development and well-being.[121] Nor after being born; for how long could a baby live without the attention, food, clothing, and shelter that others provide? An infant without a caretaker is without hope.

Starting in infancy, children learn about the world through observing, theorizing, and experimenting. They also strive toward their goals (e.g., finding and playing with a toy), reach developmental milestones (e.g., learning to walk), discover their physical and social environments, and so on. During this developmental period, they construct and fortify a foundation of knowledge about the psychological, social, and physical worlds that increases the likelihood of having predictability and control of the environment which enables goal achievement.

Yet, there's an inherent danger.

Childhood naiveté coupled with curiosity, exploration, and experimentation is sharply hazardous. Indeed, just think about common perils such as, falls, burns, electrical shocks, poisonous liquids, suffocation, choking, and drowning.

With that in mind, think about the social solution.

Children must interact with others, whose watchful eye and protective hand guard against hazards and teach safe behavior. In fact, the acquisition of most practical abilities requires social communication and collaboration.

In addition, it's during this formative period that children and adolescents begin to define themselves, others, the world, their ethical convictions and values, gender roles, social skills, goals, etc. These too are socially influenced.

Children appear to innately desire to mirror others' language, goals, actions, emotions, and rules for living. Sometimes these concepts and behaviors are aped directly from parents or peers, and give rise to phrases such as "Monkey see, monkey do."[122]

Ah, but what evidence hints that imitative behavior is genetically-inscribed?

Kids' overimitation of adults' actions can occur in infancy.[123] (Suggestively, a similar process is seen with chimpanzees who imitate human actions.[124])

Like an empty hard drive waiting for an operating system to be installed; it's as if people's genetics evolved to appreciate the value of social learning and programmed the need to acquire established goals and rules to aid survival and thriving. Childhood insecurity and self-doubt make the copy and paste of others' thoughts, rules, emotions, actions, and objectives more likely.

Children's bodily constitutions and capabilities as well as social influences and impressions shape their personalities.[125]

Moving forward, consider what's required for young adults to experience joyful camaraderie and a sense of belonging and purpose, as well as foster intimate and invigorating relationships that proliferate the species.

Socializing. The high from a person's intoxicating mixture with others serves as proof of its worth, as does the painful hangover of social withdrawal.

To acquire things such as food, clothing, and shelter, each person hones a skill or discipline. Yet, how does an individual's career have a social factor?

As complete self-sufficiency is extremely rare, individuals contribute or exchange their skills, or things related to their abilities (e.g., money) with others who have their own set of proficiencies. So, like a football team that has different players in different positions to achieve the objective of playing (and perhaps winning), complementary roles in society allow an individual to achieve goals while making the group or community better able to persist.

Cooperative effort maximizes group strengths (as each person contributes a different tool to an increasingly comprehensive toolbox) and minimizes each person's deficiencies (as others' strengths compensate). Even the housebound recluse or loner must rely like a hatchling in the nest on others to provide goods and services (e.g., Amazon.com). The examples are endless. Just think about how an orchestra can play symphonies, whereas individual musicians cannot. Or, how college students pool their money to share a big house, enjoy more conveniences, and have cash left over for pizza and video games.

Cooperation and interdependence allow for merging of resources to reach various individual and group goals.[126]

So, even in adulthood, the individual is never quite individuated.

During the latter years, seniors may enrich their friends, families, and others by circulating their accumulated wealth of wisdom and be in a better financial position for charitable acts. Also, consider how the elderly overcome any weathering in physical and/or mental capacity that may occur with advanced age through social interaction and interdependence.

Throughout each life stage an individual is, and needs to be, connected with others. Lifelong socialization is necessary, healthy, logical, and productive for an individual as well as the larger community.

Yet, maintaining interaction and collaboration is a challenge – especially so for depressed persons who deem themselves inferior and avoid vulnerability.

Thankfully, there's something that keeps people together and cooperative.

Social Interest

Those who selfishly and stealthily cheat on their spouse may lose their marriage, regular access to their children, friends, their home, and a significant portion of their money, as well as imperil their sexual health – *and* that of their betrayed spouse (as if one conjoined twin took a poisoned pill). Similarly, business leaders who take an excessive income, exploit company workers, and don't contribute to the group's success, can tarnish their reputation and bankrupt the business. This may lead you to a specific, and telling, deduction.

Individuals acting in their own interest may not only ruin things for others, but also for themselves.[127]

Life is filled with a galaxy of challenges that have a social element. Acting in a self-interested or isolated manner, regularly or extremely, is ultimately counterproductive – and jeopardizes the ability to survive and thrive.

This hints at a vital fact which has particular relevance for understanding and treating dysthymic persons who shy away from interaction and vulnerability.

Only those prepared to cooperate can tackle life's social problems.[128]

Given what you know about the need for lifelong socialization, this is unsurprising. Yet as you'll recall, *having insight doesn't guarantee that people will abide by the facts or engage in healthy acts*. For instance, people know that they should eat right, exercise, be nice, floss, etc., but alas, they often do not. So, consider what can increase the probability people will collaborate in individually and socially productive ways when facts and reason are insufficient.

Here's where it's necessary for you to become familiar with the concept of *social interest*.[129] It can be characterized as people's ability to:
- Feel at peace with the world.
- Be relaxed in others' presence.
- Freely and peaceably interchange.
- Identify and empathize with others.

- View themselves as able to meaningfully contribute.
- Feel in sync with others and have healthy social connections.
- Have a sense of belonging and see their place in a social context.
- Feel comfortable enough to share their failures as well as their successes.

First, these experiences and characteristics of social interest are ones that neurotically depressed individuals may not enjoy. After all, when people feel inadequate it's difficult for them to feel at peace, by themselves or with others, and believe they can contribute, be open, feel connected, etc. This reveals an informative diagnostic fact.

Neurotic symptoms become problematic when they persist and lead to a restriction of activity with others.[130] This is a particularly troubling paradox because depression can restrict or prohibit social interaction that would aid symptom reduction. For example, those who are reluctant to be vulnerable and express their flaws and failures would be particularly challenged to confide in others – and, believe it or not, even a licensed healthcare professional.

Second, social interest can be fostered interpersonally (among people) or intrapersonally (within an individual).

For example, a parent can encourage a child to act with others in mind, or people can decide to develop their social interest – which is a concept of use and not possession, as individuals either demonstrate it or they don't.[131]

Simply, people cannot justify avoiding healthy action by declaring they don't have social interest. This is like saying they're unable to improve their exercise level, eating habits, sexual technique, or any other changeable behavior. For those who mistakenly believe their condition is sadly permanent, *the ability to cultivate social interest creates a hopeful, empowering mindset and situation.*

Expressing genuine interest in others, having the courage to be vulnerable, accepting oneself and others, being cognitively flexible, and other displays of social interest can erode and erase depressive tendencies.

So, consider how and why extremely thoughtful and compassionate dysthymic individuals might be unable to fully experience social interest.

They may be an attentive audience and let others dominate the conversation, or repeatedly place others' needs in front of their own, for example. While this may appear to fulfill the social interest criteria…they're only partly vulnerable psychologically and emotionally – cheering from the sidelines, not on the field.

Social interest can be desirable, pleasurable, healthy, reinforced and enable connection, cooperation, and goal achievement.

"But," you may say, "if it were as simple and easy as that, life would be overflowing with social interest!" (And, if you did say that, you'd be correct.)

Like a saintly martyr, social interest can seem dangerously self-sacrificing. Indeed, socially-interested individuals risk vulnerability, loss, depression, and non-reciprocation of their kindly acts. Therefore, social interest may only go just so far in explaining social cohesion and collaboration.

Just like it's logical and healthy to eat right and exercise (which people often avoid), it's logical and healthy to act in a socially-interested way, but reason and the hope of a kind response can be insufficient to spark action.

Nevertheless, as social cooperation and exchange is a human necessity, a solution is required. So, think about what can magnify the strength of social interest as well as emphasize the importance of alliance and attachment.

Social Influence

The discussion about social interest and interdependence being healthy and necessary for human existence may seem far too ivory tower theoretical and idealistic, therefore, unlikely. After all, you are your own person, capable of making decisions, can work by yourself, and able to avoid peer pressure – plain and simple individualism in action. However, in times of doubt and uncertainty, it can be helpful to examine scientific data for illuminating evidence and clues.

With that in mind, think about what the following suggest about social influences on individual choice and functioning.

- Parental encouragement or discouragement determines whether infants crawl across a glass-covered visual cliff, where it appears to the infants that they're on a platform and any forward movement would incur a dangerous fall. So, even though the infants stop themselves, parental action encourages them to crawl over the visual cliff (they don't fall due to the transparent glass top on which they crawl). Social referencing (looking to others for clues about what's safe and suitable) may override individual decisions.[132] Social appraisal happens in adulthood as well, for instance, when a person looks to others for suggestions, which can influence emotion and cognition.[133]
- Consider how parents may promote or dissuade their children from pursuing sports, education, and certain career paths and how that can impact their kids' finances, physical and psychological well-being, social status, and the likelihood of marriage and having children. Familial factors, say parental encouragement or discouragement, can sway people's choice of various goals.[134]
- Studies show that when people are presented with three lines of clearly various lengths, approximately one-third of people tested incorrectly labeled one line as longer to conform to others' opinions (accomplices who were instructed to give the

wrong answer), or because they doubted their own senses based on others' (deliberately incorrect) responses.[135]
- Aggression can be learned by seeing another's aggressive act.[136]
- In the process of Groupthink, individuals conform to what they believe is the group consensus.[137]
- People may define themselves by how others see them (aka Looking-glass self).[138]
- Individuals tend to comply with authority figures, perhaps to a point beyond what's healthy.[139]
- Suicide rates are higher for those widowed, single, and divorced than married.[140]
- Suicide rates are higher for people without children than with children.[141]
- Children starting as young as five monitor and manage their social reputation.[142]
- News of suicides may inspire other individuals to commit "copycat suicides".[143]
- Discord with parents is an adolescent suicide risk factor.[144]
- Having more and higher quality social relationships is correlated with lower risk to health.[145]
- Those with fewer social connections have shorter life spans and are more vulnerable to a range of infectious diseases.[146]
- Those who have regular face-to-face contact with others may live significantly longer.[147]
- Those who are socially isolated may have twice the cognitive decline as those who are socially connected with friends, family, and their community.[148]
- A meta-analysis of more than 140 studies indicates that those with close ties to others significantly decrease the probability of dying over a given period.[149]
- Isolation from a significant other is associated with poor immune functioning.[150]
- Those who are lonely have weakened immune systems.[151]

- Social stressors have a more harmful effect on people's immune systems than non-interpersonal stressors.[152]
- Loneliness and social isolation increase the risk of heart disease, heart attack, and stroke compared with those who have strong social support.[153]
- Loneliness and social isolation increase mortality risk.[154]
- A spouse's death may be the most stressful experience in a person's life.[155] Revealingly, six of the top ten stressors have a social component (the remainder are: divorce, separation, death of a close family member, marriage, marital reconciliation).
- Low job status is associated with an increased probability of a heart attack than high job status.[156]
- A person's job status is a more accurate predictor of a heart attack than smoking, obesity, or high blood pressure.[157]
- The surrounding culture influences an individual's self-concept.[158] For instance, those in Eastern cultures describe themselves in more social terms with a self-image including being part of the group, whereas people in Western cultures hold more individualistic memories and self-definitions.
- Divorce can significantly increase the risk of cancer, diabetes, and heart disease.[159]
- Happier and healthier individuals report strong interpersonal relationships, whereas isolated individuals tend to experience a decline in physical and mental health as they age.[160]
- Social relationships can affect the life span on par with such factors as obesity, cigarette smoking, hypertension, and physical activity.[161]
- Increased social support has been linked to a reduced cardiovascular stress response.[162]
- Human touch and social contact can stimulate the release of oxytocin, a hormone that can reduce stress and stimulate healing.[163]
- Holding hands with someone else, synchronizes breathing as well as brain wave patterns and reduces pain.[164]

- When empathic partners touch their lover, their breathing and heart rates synchronize and pain level decreases.[165]
- The mere expectation of having sex can stimulate men's beard growth.[166]
- How an individual supports a spouse (positive or ambivalent [both helpful and upsetting]) impacts both individuals' risk of heart disease with coronary-artery calcification, with more supportive behaviors associated with higher cardiovascular health and ambivalence linked to increased calcification.[167]
- Hearing her baby cry – or just thinking about it – may make a breastfeeding mother lactate.[168]
- Mother's breastmilk changes to accommodate the nursing infant's sex, age, illness and need for protective antibodies.[169]
- Within their first hour of life, newborns can mimic others' facial expressions.[170]
- The socially-related variables of shame or embarrassment can lead to blushing.[171]
- Television comedies often use a laugh track (a recording of other people laughing) to induce individual viewers to laugh like those in the "audience".[172] (Interestingly, this isn't a new phenomenon. The French word "claque" meant a "Group of spectators receiving financial reward to applaud frantically on the first night of a new play or show."[173])
- Loneliness among the elderly is a significant health concern and leads to greater use of health care physicians and facilities.[174]
- Those with supportive spouses are more likely to take on potentially rewarding challenges, and for those who do, they experience greater happiness, psychological well-being, and healthier relationships.[175]
- Having a happy partner/spouse may boost one's physical health.[176]
- People tend to date those they believe have similar attractiveness to themselves.[177] Those who perceive themselves

as less physically attractive will accept less physically attractive individuals to date, but do not deceive themselves into perceiving those they date as being more physically attractive than they actually are.[178] (Think about how this impacts those with low self-esteem and poor self-image.)
- Limited social support is associated with an increase in cell aging.[179]
- Babies cry in distinctive ways (melodies or accents) based on learned cultural influences they listen to while they were fetuses in the womb.[180]
- Those with psychosis who hear hallucinatory voices, it's their encompassing, local culture that guides what the voices say (e.g., harsh, threatening, playful, comforting).[181]

In light of the above, you may have concluded the following.

The social environment powerfully influences an individual's health, thinking, emotion, and behavior in various ways, from microscopic functioning to large-scale social interactions.

A group, community, congregation, society, or national mood – as well as solitude – can sway a person's actions and mindset, such as language, clothing, concept of family, eating and drinking, morals, work ethic, web sites visited, sexual practices, participating in a sports stadium ritual, and personal hygiene.

Also, think about how the social environment can guide even what a person does alone when those actions have social consequences (e.g., the importance placed on getting "likes" or subscriptions to online postings uploaded from the stark, solitary confinement of a bedroom or basement.)

But there's another point to consider.

As science becomes better equipped to handle physical illnesses and people are increasingly able to acquire food, clothing, and shelter, it appears as though social stressors may supplant physical ones, or at least increase their relative proportion.[182]

For example, a spouse's death may be more troubling – and associated with various physical and emotional symptoms (including suicide) – than many physical illnesses.

Like evaluating a player on the field, given the direct and indirect social influences on functioning, a person can only be fully comprehended when seen in relation to the social environment.

A person is never just an individual, but a *social* individual.

To comprehend and conquer dysthymia, there has to be an assessment of the symptomatic person's social context and influences.

With that in mind, consider what social experience may be the most painful for an individual to endure and relevant to symptom development.

Nothing Hurts More Than Love Lost

Please take a moment and recall a specific event or experience that you consider the happiest time in your life and visualize all the warm and exciting details. Once that experience is clearly in mind, answer the following:
- Where were you?
- What was happening?
- What time of day was it?
- Who was there?

Now, bring to mind the saddest, most devastating and inconsolable episode you've ever endured, and answer the same questions.

Consider what these recollections and their associated emotions reveal.

First, recall how eating and having sex can be orgasmically pleasurable. So, unsurprisingly, people like to do both regularly. And, of course, they're the mandatory, non-negotiable entry fees to compete for the goals of survival and thriving.

Second, recollect how bumping your arm is considerably less painful than a heart attack.

And, in general, having a feedback system in which an increase in the degree of pain is correlated to the severity of the issue *allows people to rank concerns and prioritize resolving that which has intense pain*. For example, the suffering from a broken arm can forcefully stop people from working and prompt an immediate trip to the hospital.

Moreover, it can guide future behavior (e.g., "I'm never working on the roof again!"). This innate capacity also assists with survival and thriving.

Third, consider what the following medical conditions have in common.

Esophageal cancer, brain tumor, hypertension, diabetes, vascular disease, brain aneurysms, kidney disease, various autoimmune diseases, heart arrhythmias and defects, and most lung diseases.

They're *silent killers* – deadly ailments that can approach stealthily, without any pain that would warn of their attack. People don't want to be tragically unaware and caught off guard by a grave illness they'd be able to avoid otherwise. Accordingly, they can fear painless, lethal conditions as intensely as a single-minded murderer noiselessly hiding in the bedroom closet.

Now, think about what the above suggests.

First, the body has feedback mechanisms that reinforce and punish actions to guide behavior and assist survival and thriving.

Second, there's often a positive correlation such that the greater importance the behavior is to survival and thriving, there's a corresponding increase in the intensity of the reinforcement or the punishment. For example, pleasure reinforces eating and having sex, whereas agony punishes injury, broken bones, dehydration, and starvation.

Third, not having any feedback can be unnervingly dangerous, perhaps fatal.

With the above in mind, in your recollection of the happiest time or experience in your life, consider social factors (e.g., being in love, seeing the birth of a child, or agreeing with and accepted by valued others).

During your most devastating and inconsolable time, was there a *different, but related social aspect*, such as ridicule, rejection, or abandonment? For example, was your sorrow related to fight with a friend, a broken heart, or being ignored? Did you fail in front of others (or would they soon know about it)? How much would you like to go back in time and change the situation?

This may lead you to a telling deduction.

The immense, insufferable, and far-reaching agony of being ridiculed, rejected, or abandoned, reflects the extent to which people value social connections as well as underscore their importance to human existence. Note as well that hurt associated with failure verifies the importance of goal achievement and competency.

To borrow from Sigmund Freud, "It is that we are never so defenceless [sic] against suffering as when we love, never so helplessly unhappy as when we have lost our loved object or its love."[183]

To the degree that people can experience the exhilaration and narcotic ecstasy of love and connectedness, they may also endure a severe, relentless, and inoperable pain after failure or being ridiculed, rejected, or abandoned.

Indeed, there may be nothing more excruciating than a broken heart whether it's from a shattered relationship, the death of a loved one, a rejecting parent, betraying confidant, or spurning friend.

Social connection is vital, as evidenced by its intense joy, as well as by the sharp and lingering agony when it doesn't exist or is lost.

It's a pain that people may try to avoid at all cost – especially if previously experienced. Consequently, you may be tempted to say that unpleasant social conditions and outcomes can easily trigger depression.

The next chapter is a summary of hints, facts, and research that may help you realize something else that relates to dysthymia.

Notebook of Clues and Evidence

In decoding persistent depression, you've navigated around popular and seductive distractions, dead ends, and dark alleys. Also, you may have – perhaps reluctantly – surrendered previously held, but misleading and mistaken, concepts. Yet, being introduced to a massive crowd of terms, ideas, hints, and research, can be as dizzyingly overwhelming as an unruly Black Friday sale.

So, if you desire, here's an orderly parade of points (without mystery-ruining spoilers that reveal and connect every hidden clue).

1. Depression is a common and incapacitating concern.
2. The cause and effect mindset of depression is long-standing, influential, and productive, but has some risks, dangers, and doubts.
3. The way in which people conceive of psychological symptoms guides their assessment and treatment actions – even to an unhealthy outcome.
4. Rather than uncritically and completely accept any theory, assessment, or treatment, it's better to remain vigilant for any shortcomings.
5. People may mistakenly perceive their memory as an inexhaustible, well-organized warehouse of accurate facts and records.
6. Human memory is thematic, subjective, and flexible, and can guide action and protect people in present and future challenges.
7. While insight can be helpful, it doesn't guarantee symptom change.
8. Although genetic causes, or predispositions, to depression exist, they can have minimal influence, be vague, related to several diagnoses, open to interpretation, and be subject to environmental influences.
9. Reducing the complexity of dysthymic depression to a solely genetic basis may be insufficient and potentially problematic.

10. Antidepressant medications can reduce symptoms and using them in conjunction with psychotherapy may magnify the positive results.
11. Depressed people can have different levels of neurotransmitters and neuronal activity than non-depressed persons, but dwelling on depressing things may influence neuronal activity and neurotransmitter levels.
12. Psychotherapy treats people's way of thinking, which may impact neuronal activity and neurotransmitter levels.
13. The perspective that a chemical imbalance causes depression may not be universally applicable, therefore, insufficient for full comprehension.
14. The cause and effect mindset can prime people to search for a source of symptoms – and may lead to finding one, whether or not it's accurate.
15. Inattentional blindness may distract people from considering various non-causal factors that can explain depression.
16. The cause and effect narrative of dysthymia has certain plot holes, contradictions, and red herrings that may prime people to perceive a cause of depression even if it doesn't exist. (Fortunately, clues exist elsewhere.)
17. Biological (physical) symptoms (e.g., sweating, vomiting, diarrhea, fever, pain) can serve various purposes that aid survival and thriving.
18. Biological symptoms can act as feedback that identifies a threat to survival and thriving, and helpfully influence a person's movement.
19. An individual's biological symptoms can impact others' movement that benefits the person and perhaps others as well.
20. Biological symptoms can arise instantly, unconsciously or consciously, be intricately and intriguingly choreographed, have a social component, and employed to realize various goals, most notably survival and thriving.

21. Given that biological symptoms can beneficially provide feedback and guide behavior, there are times when medicating biological symptoms can be counterproductive.
22. When people medicate their psychologically-based symptoms, they may retreat from that they wish to avoid (e.g., therapy, changing their thinking or acting), yet believe they're addressing the cause.
23. Similar to physical symptoms, psychological symptoms can identify when something's wrong, be a call to improve thoughts, actions, and situations, and productively guide assessment and treatment.
24. Antidepressants may anesthetize symptoms that can resurface when the reason for their existence goes unresolved.
25. People may mistakenly believe their psychotropic medications cured their depression when it's due to other factors that have changed.
26. When medication alleviates psychologically-based symptoms, people may fortify a misbelief of an innate, organic symptom origin.
27. As psychological symptoms can provide profound and helpful clues about people's cognition, perspective, and goals, there are times when merely suppressing them medically can be counterproductive.
28. The growth model avoids many of the difficulties and addresses the drawbacks and uncertainties of the cause and effect mindset of symptoms.
29. The growth model views symptoms as a function of development; for instance, improper or insufficient learning that can be remedied by studying the thoughts and behavior of non-symptomatic individuals.
30. Growth model therapists view people as active, creative, and social decision makers who pursue a range of desires. They

focus and build on people's strengths and encourage healthier thoughts, behaviors, and goals.
31. Similar to biological symptoms, an individual's depressive symptoms can influence others and outcomes in the social environment.
32. Goals are vitally important in human functioning.
33. Humanity's top two goals are to survive and thrive.
34. The life tasks are areas of functioning that lack inborn answers and which people must address by selecting goals and how to achieve them.
35. People can be strongly motivated to acquire skills to compensate for insufficient innate knowledge of the world and how to act within it.
36. Some attempts to overcome inability can be counterproductive.
37. Individuals may experience inferiority feelings associated with challenges in various areas of functioning and throughout the life span.
38. Inferiority feelings can influence goal-attaining movement (e.g., avoid dangerous action, prompt learning a new skill).
39. Inferiority feelings do not cause depression, rather it's how individuals choose to handle them.
40. People's goals influence what they do with their inferiority feelings.
41. Inferiority feelings and depressive symptoms can stop people from taking goal-achieving action that's within their ability.
42. Even if people prove their capability to others, until they recognize their abilities, they can maintain counterproductive inferiority feelings.
43. Confidence is realized only after reaching a goal and proving ability to oneself, until then there's just a desire and the drive to attain it.
44. People must be competent in numerous ways to survive and thrive.

45. Demonstrating competency can have a social factor.
46. People pursue what they value and believe help them survive and thrive.
47. Competence is defined in countless ways.
48. People don't need to be the best to consider themselves significant.
49. Dysthymic persons strive for goals they believe grant competence.
50. Those with low self-esteem can fear failure and negative judgment, thus feel daunted pursuing common goals, or doing so in the usual way.
51. Lifelong socialization is necessary, healthy, logical, and productive for an individual as well as the larger community.
52. Only those prepared to cooperate can tackle life's social problems.
53. The social environment powerfully influences an individual's health, cognition, emotion, and behavior in various ways, from microscopic functioning to large-scale social interactions.
54. To comprehend and conquer dysthymia, there has to be an assessment of the symptomatic person's social context and influences.
55. Social connection is vital, as evidenced by its intense joy, as well as by the sharp and lingering agony when it doesn't exist or is lost. It's a pain that people try to avoid at all cost – especially if previously experienced.

The above hints at how and why dysthymia can rise, survive, and thrive.

Next, given the relevance of social factors in symptom development, consider what's perhaps the biggest social context a person can be compared.

Common Sense

Think about the shared elements among baseball, chess, cricket, football, poker, rugby, tennis…and most other games.

To start, each is a *competition* among contestants who *strive to win*. Players attempt to achieve this by earning one or many *subgoals*, such as scoring points.

Another similarity is that there has to be *social cooperation*. This usually occurs in two forms. The first is among teammates who work together to achieve the main objective of winning. The second occurs between competitors who agree to interact to achieve their shared aim of playing the game. These elements reflect the importance of goal setting, achievement, and collaboration.

Ah, but there's an additional, indispensable element of gameplay.

Given that participants cooperatively interact to achieve various objectives, imagine the disorder that would happen without general guidelines of acceptable action.

It would be a mess.

There would be exasperating uncertainty about goals and the means to attain them, with some acting in self-interested and seemingly random ways that jeopardize cooperation and goal achievement. The participants would become frustrated, confused, and angry. This counterproductive inevitability eventually sabotages the overarching goal by making it impossible to play the game.

Now consider how the solution requires social cooperation.

Through communication, officials and participants come to a *consensus* about the *common rules* they must know and *adhere* to so they can play fairly and coherently. And if they don't, penalties are given.

They can be officially dispensed (e.g., players temporarily prohibited from participating, or they're ejected from the game). Or, like siblings that shush the misbehaving one who invites a parent's untimely wrath, the coach or teammates reprimand the rule-violating player who endangered the team's ability to win.

General rules that everyone accepts are mandatory to play a game.

There are parallels between gaming and life.

Both have ultimate goals.

In gaming, it's to win. In life, the biggest objectives are to *survive* and *thrive*. In addition, recognize that each endeavor has *subgoals* that must be achieved to attain the main goal. In games, players must earn points. In life, there are many subgoals that serve survival and thriving (e.g., earn an income, eat, find shelter).

Also, consider how some subgoals are related, but optional (e.g., get married, have sex, raise children). Note as well the requirement of lifelong socialization, interaction, and cooperation.

So far, so good, but also appreciate the following.

People have an unavoidable need to understand life (e.g., others, how things function, which foods are safe to eat) and control those things that lead to goal achievement (e.g., tools, speech, muscles, machinery). However, life doesn't come with a set of instructions. Therefore, imagine what can arise from a lack of common and consistent ways of thinking and acting. For example, if people had dissimilar conventions for social interactions and violated common standards and perceptions when in a movie theater, courtroom, public restroom, restaurant, business meeting, or bank it would be a mess.

The solution is similar to that for games.

When people don't understand how certain objects or environments work, they may theorize about what's right, true, and advantageous.

Through communication and consensus among society members, each community cultivates a fluid pool of common knowledge of generally accepted and encouraged rules and goals that evolve over time.[184]

For instance:
- Roles in the environment (e.g., children go to school, adults go to work).
- Rules for driving, socializing, working, living, and dating.
- Language acquisition and usage.
- Common goals and how to achieve them, for instance, to find a job, get married, have children, save for retirement.

- Mutually agreed upon concepts, such as gender roles, religious beliefs and practices, health and safety issues, ethical convictions.

Social cooperation allows groups, communities, and countries to create and share guidelines of acceptable action and objectives that are necessary for coherent interaction as well as individual and group goal achievement.

Common knowledge, aka *common sense*, is essentially a mutually agreed upon set of principles, beliefs, and aims, often arrived at by consensus, and seen as realistic, proper, accurate, and healthy, when the majority of people in a group or subgroup (e.g., family, congregation, constituents) consent to them.

This leads you to a mind-blowing conclusion.

Largely, the rules and perceptions by which people live are subjective human creations projected onto reality and acted upon as if they were true.

Nevertheless, like a playing a game, people can function well when they act in accord with generally-accepted concepts, rules, and goals. For example, you believe that the government-issued cloth and circular pieces of metal in your pocket are of value and freely exchange them for goods and services with others who also believe those items have worth. Alas, like a game or shaky marriage, expect that some people will shatter agreements, break the rules, and cheat.

There's something else that shows the significance of common beliefs, actions, emotions, goals, perceptions, logic, and values.

Perhaps you've experienced a certain sense of joy, connection, compatibility, or peace when others know and say the same catchphrases or movie quotes, maintain the same political, scientific, or religious perspective, or follow the same sports team.

This uniquely pleasurable feeling reveals and reinforces the importance of a shared reality, bonding, and cooperation born from common knowledge.

For people to address life's challenges, they communicate and cooperate to create and improve their perceptions, goals, and rules for living. They can act coherently and the positive feelings associated with kindred cognitions and desires reinforces the process and testifies to the usefulness of common sense.

Next, consider how the following regularly occurring conditions provide evidence that living by shared guidelines and common knowledge is important on a larger scale.

- Praise and recognition for good sportsmanship.
- Tax breaks for being married and having children, owning an energy efficient car or home, and for donating to charities.
- Dean's list and honor societies for those who achieve high marks in academics.
- Awards and bonuses for performing well at work.
- Military medals and ribbons for heroism and distinguished performance.

Groups, societies, businesses, and nations reinforce compliance to commonly-held and valued goals, actions, and ways of thinking.

As if the preceding weren't enough, consider what else reveals the importance of shared perceptions of reality, common rules of thought and action, and socially-acceptable goals. To start, think about what occurs when players violate game regulations.

Recognize the importance of social conventions, say at a party, within the workplace, or in public, and how people can become subtly to noticeably tense when those guidelines are violated, and how they may offer unsolicited advice, or physically intervene (e.g., pulling someone aside to talk, or guiding them toward the door and away from conflict) to calm the waters.

Disobedience of common perceptions, guidelines, goals, and laws is often met with social as well as official censure.

For example, people may be the subject of litigation, imprisonment, disapproving looks, direct statements, harsh tones of voice, among other forms of feedback (including a quick elbow to the ribs from a loved one).

Or, consider how in virtually every country around the world it's accepted, encouraged, and authorized that societies have law enforcement agencies in place (e.g., police, judicial systems, legislative bodies) to maintain public order, ensure safety, and prevent crimes, by forcefully compelling compliance to the general regulations of thought and action.

The populace gives these agencies the authority to deem individuals a threat to the common knowledge and social order, and can deprive them of their money and freedom. This can protect and serve others who abide by the shared laws (e.g., decrease the probability of robbery, rape, and murder).

Customarily, citizens grant law enforcement agencies ultimate power over every individual in a society – the ability to impose supreme punishment to enforce compliance to the public regulations in a principled and worthwhile attempt to guard the law-abiding residents from harm.

Some societies cherish their common sense to such a degree that they kill their own inhabitants who violate the common standards and goals.

This testifies to the value people place on protecting and perpetuating...

- Shared, socially-acceptable perceptions of reality.
- Cooperation.
- Common rules and objectives.
- Individual and group goal achievement, including survival and thriving.

And notice that the offenders' punishment usually imperils or ends their ability to thrive or survive (e.g., fines, imprisonment, asset forfeiture, capital punishment). This underscores those goals' importance, as well as a community's or nation's need to serve and protect the majority of its populace.

Next, think about another example of socially-permitted and encouraged compliance to shared perceptions, canons, opinions, actions, and objectives.

Consider how some religions compel their followers' obedience by warning that they'll suffer for disobedience, whether in their present life (e.g., shunning, excommunication, stoning to death), or in some form of infernal hereafter.

Can there be attempts to get those of a *different* group or nation to comply with shared religious dogma, mandates, cognitions, and customs?

Sure, just think about present and historical examples of forced conversion, religious persecution, pogroms, sanctified violence, and murder.

So, what's perhaps the most extreme illustration of imposed obedience to shared perceptions of reality, guidelines, and goals?

War.

Sometimes it's seen as large-scale armed robbery of valuable assets. More positively, there are times when armed conflict can justifiably enable nations to protect their citizens against murderous attack. Or, like a watchful neighbor stopping a bully from beating up a weaker innocent kid, military intervention has altruistically and beneficially enforced human rights and protected those living in different countries from exploitation (e.g., genocide, war crimes, crimes against humanity, ethnic cleansing).

Yet, wars' epic intensity and severe consequences show the importance people place on having shared conceptions of reality as well as common rules and goals.

By imprisonment, financial and physical punishment, and death, some countries attempt to enforce their perception of reality, rules, and objectives on people in other nations. There are graveyards around the globe filled with those who didn't share, and refused to adopt, the invading group's common sense.

Next, consider why wars are often based on religious or political perspectives.

It might be, in part, because religions and governments are the two largest groups that create and maintain specifically-defined common knowledge.

Nevertheless, the point is that wars can forcefully perpetuate a society's common sense by getting others to surrender their own.

NOTE: *This isn't a polemic for or against war or any political or religious system.* Rather, real-life examples illustrate the degree humanity values reality-defining common knowledge and lengths taken to institute it broadly.

The countless settings and ways people enforce compliance to common sense reveals the immense value they place on having similar perceptions of reality, agreement in thinking, acting, and aims, as well as cooperation to attain goals.

Like a teacher trying to get all of the students to behave so that everyone can enjoy the field trip; in general, groups and societies want everyone to act harmoniously according to the same rules to reach socially-valued goals.

Now, consider what this detailed discussion about common sense has to do with depression.

Common Sense May Define Mental Wellness

Common sense is the prevailing knowledge within a group, community, or nation that's usually arrived at by general consensus and viewed as most in line with reality and appropriate.

Accordingly, those whose perception of reality and rules of living embody what each society values (having the "best fit" to the general trend) are deemed as normal and healthy, and apt to be acknowledged, accepted, and praised. For instance, consider those whose parenting skills exemplify the common knowledge and are thus deemed good, healthy, right, desirable, and realistic.

Obversely, people may perceive those who don't act in accord with common sense as mentally ill, odd, or some other critical assessment. They can become frustrated, confused, or shocked by others who "just don't have any common sense!" For example, people may think negatively about those who drive at a significantly different speed, talk in the library or movie theater, or leave their shopping cart in the middle of the aisle, which blocks others from getting by.

Now, consider how dysthymic individuals may feel awkward, painfully different, alone, alienated, unlovable, and as hopeless failures when they perceive themselves as thinking and acting outside of common sense.

Recognize that the encompassing *community* may influentially define an *individual's* mental status, with thoughts, actions, and feelings that deviate further from the common knowledge deemed symptomatic.

Laypersons may use common sense as a yardstick to assess mental health.

However, there's an inherent drawback with this tendency.

Problematic Common Sense

As people have consuming responsibilities (e.g., parenting, working, paying bills), some may rely on easy-to-remember, but misleadingly simplistic popular beliefs (e.g., it fits on a bumper sticker) to understand things. For instance, some make an unqualified depression diagnosis based on one act (e.g., "He gave me his old football. He must be depressed") as it's commonly stated that if someone gives things away it's a sign of depression, possibly suicidality.

Next, consider what the following demonstrate.
- Myths
- Stereotypes
- Superstitions
- Folk remedies
- Urban legends
- Old wives' tales
- Exploitive religions and political systems

Even with communication and consensus, there are times when accepted and advocated common sense can be inaccurate, counterproductive, or unhealthy.

People may project false and hazardous beliefs onto reality and act as if they were true, healthy, etc. They may encourage and implement harmful guidelines and goals that ultimately imperil individual and group goal achievement. Just call to mind present or past cultures in which slavery is accepted, overeating is encouraged, genocide is advocated, women are denied equal rights, and child labor, neglect, or abuse is tolerated or condoned.

But that isn't the only troublesome factor.

Given that common knowledge represents what a community or culture values and agrees to be appropriate…

- Would common sense be the same among places such as Beijing China, Mineral Point Wisconsin, or within a modest African tribe?
- Is common sense consistent between cultures and subcultures, for example, between Republicans and Democrats, Southern Baptists and atheists, families that value athletics and those that dismiss sports in favor of academics?
- Does common sense remain constant throughout time?

Of course not.

Given the variations across time, place, and culture: *Common sense isn't very common.* For example, imagine traveling into the past and telling people that there isn't a god that brings the sun across the sky, but in actuality the earth goes around the sun, or that human sacrifices do not guarantee good health, but that germs and viruses (among other things) cause disease.

Given that your beliefs wouldn't be in accord with common sense, you may have been considered a lunatic, as the common knowledge at one point in history was that changes in the moon caused eccentricities.[185] What's more, you may have suffered socially-condoned and enforced shunning, physical punishment, or worse – all based on now-discredited beliefs.

Examine the evidence:
1. Largely, communication and consensus create common sense.
2. Common sense is subjective and varies over time, place, and culture.
3. There are times when common sense is inaccurate or unhealthy.
4. People, groups, and societies tend to promote and enforce compliance to common sense and may punish those who violate shared values, goals, and guidelines believed to be true or good – even when they're not.
5. Laypersons may reach incorrect conclusions of others' behavior that's discordant with common sense.

6. Common sense may incorrectly define thoughts and behavior as illness.

Viewing people through a distorted and variable common sense lens has certain implications for defining mental wellness, as well as comprehending, and treating psychological symptoms.

Social representations are theories, values, beliefs, metaphors, notions, etc. held by a group, which are created and operate at a social level within common sense, and used to understand, communicate, act, and explain, including those with psychological symptoms.[186]

For example, laypersons may rely on social representations of scientific theories for depression: chemical imbalance, trauma, genetically "hardwired", etc. But, like someone who recorded a live concert on a phone and uploaded it to the Internet, social representations may not have the fidelity to the original, but are widely used.

Though laypeople are often without sufficient awareness, knowledge, or clinical experience…some can be supremely confident and seemingly as immovable as Mount Everest in their incorrect diagnostic verdict.

Individuals or groups may arrive at mistaken conclusions when they hold inaccurate or factually-incomplete common sense.

But don't worry, communication and consensus can be a productive path to increasingly accurate and healthy common sense. Knowing that, consider what may reduce dangerous misperceptions and practices based on incorrect popular knowledge (including how people think about depression).

Common Sense Solutions

Every nation has insular regions teeming with insufficient or inaccurate common sense, seemingly as impenetrable to moral philosophy and science as a barroom brawl.

Yet, it's a resolvable problem and biology offers the first clue.

Human genetic diversity, though less than other species, has fortified people's defenses against environmental threats (e.g., infections, diseases).[187] That is, individual genetic mutations that improved functioning and became widely-shared have made humanity in general better able to survive and thrive.

Consider how this applies to minimizing problematic common sense.

Just think of a person who invented a tool (e.g., fork, compass, magnifying lens, scissors, rope), that once distributed advanced countless populations.

What if such individual contributions could be multiplied and disseminated?

For instance, consider the immense brain power available in crowdsourcing and cooperative learning, which can accelerate the accumulation of information and rapidly propel humanity forward as if powered by countless rockets.

Multiple perspectives can provide unique and novel solutions as well as prevent replicating narrow and unsound beliefs and practices. Just think about the global benefits from sharing medical, technological, and engineering advances, or how a small town can offer international forms of food and music.

Diversity in thought, viewpoints, goals, and actions can benefit humanity in countless ways making it healthier and more robust.

But there's a few high fences that need to be overcome.

First, consider how can there be unrestricted and omnidirectional exchange of ideas and practices (say, from a bus-riding patent clerk to the hallowed halls of the most advanced universities, or vice versa).

Unobstructed availability of information and global dialogue. For instance, while there are many areas without Internet access, people around the globe increasingly have instant access to information. This can allow a means to contribute, access to accurate and healthy concepts and practices, as well as reduce groupthink, cultural lag, and limiting perspectives.

Second, a herd mentality (emotional, non-rational discussion) doesn't have to stampede and overrun logic, as is seen in well-moderated online forums that keep the discussion positive and reasonable.

Third, critical thinking and rational skepticism can be used to exhaustively scrutinize and test the validity of a throng of perspectives, keeping the most accurate. This enables various forms of common sense to converge and fall increasingly in step with reality, and prevent misinformation, inaccurate beliefs, and myths from flooding the public square like vacationing revelers on holiday.

Now think about how this relates to dysthymia.

Common sense influences how people define, consider, and handle psychological symptoms. Accuracy is achieved through communication as well as skeptical inquiry during the process of consensus. Information is discussed, tested, revised if necessary, implemented, tested again, and so on, in a principled and virtuous attempt to continually improve understanding and care.

This explains, in part, why there have been multiple versions of the diagnostic manuals employed in psychology and psychiatry (Diagnostic and Statistical Manual of Mental Disorders [DSM] and International Statistical Classification of Diseases and Related Health Problems [ICD]).

Simply, they evolved over time, with different diagnoses and definitions to accommodate the ever-increasing knowledge of psychological conditions. New diagnoses were created, others maintained or refined, and some went extinct.

In regard to depression, think of a classroom full of students.

Those who are popular, vocal, productively contribute, and whose familiar older siblings with a well-liked reputation may have paved the way for easier acceptance, represent the cause and effect mindset.

And, you may liken the growth model to the quiet, less popular students that provide contrasting but valuable perspectives that reasonably challenge prevailing common sense or social representations of psychological symptoms, thus broaden group comprehension.

For instance, rather than view those with dysthymia only as passive victims of their genetics, biochemistry, social environment, work, relationship, or other influence, or overemphasizing and being too reliant on memory, insight, and medical interventions, *it's beneficial to understand how people are active, creative, social, make choices, and move toward a range of objectives.*

Self-determination

Recall the example of how people act differently whether they're in a creepy dark alley, an amusement park, their living rooms, a business meeting, a rowdy sporting event, a solemn courtroom, or an insufferable relative's house.

While this can initially appear as though the social environment dictates people's behavior, *their actions are based on selecting and striving toward goals in each of those settings.*

Individuals change their presentation and behavior in those situations *to achieve their objectives* (e.g., sidestep getting mugged in a dark alley, fully experience the joy of the park, relax and be free from social judgment within their living room, avoid a long and gloomy wait in the unemployment line, guard against losing their court case, prevent chaotic family conflict).

Although the social environment is influential, individual choice can be the most important factor in goal selection and achievement.

For instance, how do you explain siblings who were born and raised in the same house, yet choose different careers, hobbies, and social groups?

Sure, they may celebrate the same holidays and share numerous customs, but they can be fairly distinct. Essentially, they exercise their autonomy by selecting diverse goals and the means to achieve them. Despite siblings' genetic and upbringing similarities they still choose whether to accept or reject what's presented. (Think of it this way: While in the same restaurant they get identical menus, but can choose different entrées.) Honestly, do your siblings act, think, and have the same interests and goals as you?

Probably not. After all, each of you has individual choice.

"But," you may ask, "what about *identical twins*?"

Believe it or not, identical twins can be quite distinguishable in character and desires. This reinforces the importance of self-determination (and implies much about astrology). Indeed, in what's known as the "Teeter-totter effect" different siblings – twins or not – may select goals and succeed in different areas, as they tend to seek competence in their own way.[188]

If you're curious about how early self-determination appears, consider what can happen if you were to try to coax a two-year-old to take a nap, stop watching TV, eat vegetables, or do anything the little one doesn't want to do.

The otherwise adorable and charming tyke may plead, kick, scream, throw things, cry, run away, withdraw, shut down, or do whatever else comes to mind. These forms of resistance provide substantial (and perhaps exasperatingly ear-splitting) evidence that the child not only has a different desire than you have, but also is rather tenacious.

Even at a very early age, self-determination in goal selection and the behavior to achieve it is easily detectable.

Now consider what can testify to its strength throughout the life span.

Perhaps you've experienced how grueling it can be to talk people out of being in love with those who are bad for them or get others to stop smoking or drinking to excess even if they're fully cognizant of the injurious effects.

Sometimes, a transfusion of reason may not inoculate against, but intensify a dangerous situation.

People's stubbornness in their desires and the means to achieve them – even when it's harmful – *demonstrates how free will can override innate goals such as health*, to the point where they become immune to argument.

Individuals must make decisions and choose their behaviors, often as a means to become competent in their environment.[189] Perhaps most revealingly, *when people feel ineffective or experience a loss of self-determination they may feel helpless and become physically ill*.[190]

Accordingly, it may be easy for you to see how the loss of individual choice could cause depression. But also consider how else can the two be related.

Self-determination doesn't mean that people like to be in charge all the time or that they must control others.

Just take a moment to consider how even those individuals who want control choose to relinquish it when it's tied to weighty responsibility or punishment, when they're indifferent to the choices, or when there's slight or no chance for successful goal achievement. For example, it may be less stressful for football players to sit on the bench than to be on the field and held accountable for losing the game. Or, think about how it can be safer to have someone else be in charge of choosing the movie or restaurant rather than being blamed for making a bad choice and have others be displeased.

Therefore, having the perspective that self-determination means that people must be in control doesn't take a large enough view of free will. *The biggest choice of all is whether or not to be in control in the first place.*

Ultimately, it may be less stressful to let someone else take the wheel. For example, during a long and difficult road trip it may be easier to take a nap in the backseat than to be in the driver's seat – except for those who *really* like to be in control and couldn't sleep because they don't feel at peace with others in charge. Relatedly, some people choose to drive across the country rather than take a plane that's far faster, more convenient, and less expensive, because they're not in control of flying it.

When some people are overwhelmed or intimidated by the choices before them, they may fail to make a decision. This too is obviously a choice. Some goals can be met by relinquishing control, through indecision, or default (e.g., sitting in the back row of a cavernous classroom and not called on).

Consider what the following individual choices can reveal.
- To join the priesthood.
- To enlist in the military.
- To be social or be a loner.
- To get married or be single.
- To have children or be childfree.
- To engage in self-injurious behavior.
- To eat healthily or consume junk food.

- Good or poor dental care and bodily hygiene.
- To pursue aspirations or be intimidated by them.
- To go to graduate school or drop out of high school.

Individual choice and goal selection impact how people live, how they appear, their capability to survive and reproduce, as well as influence others.

Interestingly, there's one stone that has been deliberately left unturned.

Even though all the previous examples show how individuals move toward their objectives, that doesn't mean that action is necessary for goal achievement.

Surely, you or people you know had the desire and chose to "do nothing" while on vacation or during a lazy Sunday. Undeniably, *people can achieve goals through inaction*.

Numerous factors shape goal selection and achievement, which impact the quality and quantity of life, as well as underscore how people are goal oriented and must be that way. Each individual does not simply consume and regurgitate society's common knowledge. Rather, every person contemplates and samples from society's buffet of perceptions, rules, and objectives selecting those most appetizing, with some guidelines and goals improvised or created.

Yet, you may remain skeptical of the importance of self-determination as well as human goal orientation and desire more evidence.

Fair enough. But consider how 100% of your direct-line ancestors ate and had sex. This is an indisputable truth as you wouldn't be here if either of those actions did not occur. Accordingly, the genes that prompt eating and sex were strengthened and kept in the gene pool. And, you're the one who is currently swimming through history.

While you're well aware of your own desire for food and sex, consider how there are individuals who *voluntarily* refrain from either or both *to achieve some desired outcome* (e.g., weight loss, avoid disease and maintain a healthy body, protect their reputation).

Individual choice in goal selection and achievement can override some innate, genetic predispositions (e.g., people may refrain from getting angry, choose to have plenty of sex but not any children, overeat despite their body's feedback mechanism to stop, jump from perilously high places in the face of massive fear that could intimidate a hostage negotiator).

Your increasing realization of the role of self-determination allows you to further discount a purely cause and effect explanation of depressive symptoms.

Now, consider where the importance of individual choice – especially in comparison to genetic, physiological, or biochemical factors – may be most pronounced and relevant to demystifying dysthymic depression.

Psychiatry and Psychology: Hardware and Software

People may incorrectly use the related words psychology and psychiatry interchangeably. Unsurprisingly, there also may be some confusion about the definitions and differences between psychologists and psychiatrists.

Psychiatry and psychology are more like cousins than identical twins. Each has a perspective and unique role in addressing people's symptoms. Yet, they can be employed cooperatively and enhance each other's contribution to the process. To clarify terms and functions, consider the following analogy.

Think about how your phone has hardware; those tangible, physical components, such as a screen, central processing unit (CPU), battery, buttons, antennae, microphones, and speakers.

If you dropped your phone, got it wet, or there's some issue with the wiring or flaw in a physical component, then you have a *hardware* issue that causes various symptoms (e.g., the phone doesn't turn on, or it freezes, crashes, or hangs). The logical action is to address the hardware problem with a hardware solution; have the phone fixed or replaced.

Also, consider how your phone also has *software*; the operating system and apps, which are composed of intangible lines of computer code.

Essentially, software contains the rules, commands, data, and goals that:

1. *Create the reality of each device* (i.e., how it presents in sight, sound, and touch, its directions for functioning, what objectives are allowed).
2. *Tell the otherwise dormant hardware what to do* (e.g., search for an address, play music, send a text).

The software activates and controls specific hardware to achieve a goal. For example, to take a picture you start an app that wakes and runs the camera.

It's interesting and necessary for you to note that the CPU (the phone's "brain") – as well as other hardware – mindlessly and mechanically does only what the software directs it to do. It *must* function within the software guidelines set by the programmers and it *must* submissively obey your commands to achieve certain goals (e.g., play a video, call your friend, start a game).

The software can control what the hardware does. That is, the hardware can only function within the parameters set by the software. Also recognize that *you* are only allowed to do things that the software permits (e.g., you can select an image or play a game *only* if the app is available for you to do so).

By the same token, hardware can limit what the software (and you) can do. For instance, without a camera, software to take a picture is useless. *Hardware and software must work cooperatively for the phone to function properly.*

Usually, the software code is executed as flawlessly as a renaissance marble sculpture. Yet, ruling out any hardware issues means that crashes, freezes, and hangs are software related.[191] Simply, it's a *software* problem that's responsible for those difficulties. Like how just a couple of cars can cause a massive traffic jam, most likely it's *a relative few lines of code that are problematic.* The good news is that the entire operating system or app doesn't need to be rewritten to resolve the symptoms, just a comparatively small portion.

There's no need to address hardware for a software concern.

Instead, your solution is to get a software update to your phone's operating system or app to address symptoms and reduce the probability of experiencing the same issues in the future. Once the software-based rules, commands, data, and goals are more accurate and appropriate, the likelihood of crashes, freezes, and hangs decreases significantly.

Consider how this analogy relates to psychiatry, psychology, and dysthymia.

Just think about how the human body has hardware (e.g., neurons, hormones, neurotransmitters). Sometimes, hardware concerns (e.g., brain injury, neuronal issues, malnourishment, infection) cause symptoms (e.g., depressed mood, decreased energy, agitation, limited concentration).

Hardware functioning largely falls within psychiatry's jurisdiction. Accordingly, depressive symptoms related to issues of the observable and tangible may be addressed psychiatrically, say medication or diet changes, that targets and improves things such as, neuronal functioning, neurotransmitter amounts, hormone levels, and nutrition. *Psychiatry can ably, productively, and compassionately treat hardware-based depressive symptoms.*

So, metaphorically akin to phones, the body has hardware, and there can be hardware-related symptoms as well as hardware-based solutions.

But people have something comparable to software.

First, just as your phone software consists of rules, commands, data, and goals, people have complex, individualized "software" of personally-selected rules of logic, commands for action, information, and objectives that guide thought, emotions, and behavior for countless matters (e.g., socializing, work ethic, morals, marriage, family, career, sex, exercise, speech, diet). Of course, people are not phones or computers, it's just an easy, illustrative analogy.

Second, similar to how an operating system and apps produce the reality of your phone (e.g., how it presents, perceives the world, functions), *each person's software constructs a unique picture and comprehension of reality.* (That *is* reality, from each individual's perspective.)

People act in accord with their software's definition of reality and the associated guidelines of thought, action, emotion, and desired objectives.

Third, acknowledge that your phone software instructs the hardware how to operate. With that in mind, consider why people do the following.

- Get up and out of bed and go to work.
- Read bedtime stories to their children.
- Run and extend their arms to catch a ball.

- The unique ways in which they participate in certain holidays.
- How they donate their time and effort to a cause they deem worthy.

People's personal software can tell their hardware what to do.

Think of this as similar to how your phone's camera turns on only after the software starts it. Of course, there are exceptions, such as involuntary reflexes and action (e.g., blinking, heart rate, blood pressure) as well as biological troubles (e.g., seizure, stroke, malnutrition) that impact functioning. Or, consider those who wish to be a sumo wrestler, but don't have the required skeletal build and strength. So, in what may sound very familiar, hardware can constrain software. (This may remind you of the nature versus nurture debate, in which nature and nurture are intertwined and interact with each other.[192])

Whether it's distinct areas of the brain, certain muscles, or related physical structures, people's software can call upon specific hardware to reach a goal.

Knowing this, think about what can you deduce from people's actions.

Just imagine watching someone at a grocery store lifelessly look at vegetables for a while...but not take any, then the person briskly walks to the ice cream aisle and rapidly puts several pints of ice cream in the shopping cart.

People's hardware can reveal their software.

Next, consider how individuals' heart rate increases when their sports team wins, or when they've done something awkward in front of valued others they wish to impress. Also, note that people's physical sexual arousal is generally in accord with their rules and goals (e.g., when and where they can have sex, if they're happy and attracted to the person).

The link between people's hardware and software doesn't have to be conscious or voluntary, but it is in accord with their objectives.

But this informative connection isn't only for activating action.

For example, people may be biologically-inclined to have sex regularly, but they may have rules which strictly prohibit such behavior. Or, individuals may wish to buy a sports car, go on a lavish vacation, or some other expensive endeavor that would not only be problematic financially, but also lead to chronic and insufferable marital conflict, so those actions aren't taken.

People's software can activate or suppress specific hardware.

Individuals' software doesn't control 100% of their hardware…but, it may be more than you thought.

Fourth, consider how just a few lines of inaccurate software code can hinder or stop your phone from functioning. Knowing that, think about how inadequate rules for parenting, saving money, exercising, diet, socializing, etc. can significantly imperil functioning.

A relatively few guidelines, perceptions, or objectives can slow, corrupt, or prevent people's ability to competently address certain life tasks.

Now, imagine someone moves from a small town to Los Angeles. But, the person retains the belief (think of this as a line of personal software code) that anyone can drive across Los Angeles in 10 minutes, as if crossing a small town.

In addition, to conform to this mistaken belief, the person rationalizes that even though Los Angeles is much bigger than a small town, the Los Angeles Freeway has about 20 lanes that allow an unencumbered flow, as if it were a plaque-free artery in an oatmeal-eating vegan.

So, consider what happens by maintaining this erroneous bit of software.

While driving on the Los Angeles Freeway, the person may become angry, frustrated, depressed, anxious, tearful, punch the steering wheel, or show some other expression of this minor psychological software glitch.

Note as well that there *has to be* the associated changes in how and where the brain activates or neutralizes neurons, neurotransmitters release, muscles move and tense up, heart rate and blood pressure changes, etc.

Just think about how when people are rejected or fail due to their rules, information, or perceptions, they may experience a number of physical manifestations of depression, e.g., reduced energy, crying, eating changes.

People's software can control their mood as well as their hardware.

Fifth, there's a common misbelief that intelligent individuals "should know better" or be "clever enough" to avoid depressive symptoms. But this is a misunderstanding between people's hardware and software.

Keeping with the phone metaphor, liken an intelligent person's brain to an extraordinarily-powerful central processing unit (CPU) – the phone's "brain". No matter how fast and capable the CPU, if software glitches exist, this can lead to symptom development (e.g., the phone is unable to perform a task, freezes when unable to make a decision according to the software rules).

This suggests something about those with neurotic depression.

Dysthymic individuals can be exceedingly intelligent, but that doesn't necessarily inoculate them against psychological symptoms. In other words, they can have a very powerful brain, but if their software is inefficient and/or ineffective they can experience depression.

When people's phones don't work well, they may not know if it's due to a software or hardware problem. And, given the difficulty in comprehending how software works, it may be *cognitively easier* for them to believe something is physically faulty. (Given that everyone in childhood has broken a toy, people can readily grasp the concept.) Likewise, some people default to a hardware conception of depression rather than investigate the intricate and complicated aspects of intangible rules, thoughts, emotions, information, and goals.

Software problems may look like, and be mistaken for, hardware problems.

Just like there can be human hardware issues that create depression and require psychiatric intervention, people's software can lead to symptom development that looks like a hardware issue but requires a software solution.

This metaphor may help explain psychiatry and psychology mindsets.

To keep this straightforward, the focus will be on psychiatrists and clinical psychologists who treat patients (rather than researchers, for example).

Both psychiatrists and psychologists strive to reduce people's symptoms. Moreover, there's some overlap in their training. Psychiatrists have instruction in psychological theory and treatment, and psychologists are educated in biology, psychopharmacology, and physiology. In addition, consider how some psychiatrists conduct therapy and some psychologists prescribe medication.[193]

Okay, but there are some identifying distinctions.

Psychiatry and psychology are fundamentally different disciplines with clearly-distinguishable approaches. In fact, given the similarity of training (undergraduate, medical school, and residency) psychiatrists may have more in common with *neurologists* than psychologists.[194]

Psychiatrists are medical doctors predominantly trained in the biological aspects of their patients' symptoms and may view disorders as abnormalities in physiological form or functioning (e.g., neuronal structure, neurotransmitter levels, hormones, nutritional issues), and treat symptoms by medical means.[195]

Psychiatry does an admirable job in addressing hardware-based symptoms, often by employing tangible, physical interventions.

"But," you may ask, "why can a hardware solution – say medication – reduce depressive symptoms related to people's software?"

Recall the e-mail filter and analgesic examples. Now, imagine that a software glitch sets your phone's volume and screen brightness to the maximum. Putting tape over the speakers and screen reduces the volume and brightness but doesn't fix the software issue related to those symptoms. Comparably, medication can influence people's hardware to reduce their psychologically-based symptoms, but it cannot address the software.

Psychology assesses and treats people's software.

Psychologists are doctors who have focused and rigorous training in various psychological theories and techniques. They address people's symptoms that are related to cognitive, behavioral, and emotional factors, which can be improved by implementing non-medical interventions.[196]

Like removing a pebble from your shoe that's related to psychological, physical, and social symptoms (sore foot, leg and back pain, grumpiness and social discord), therapy identifies and replaces troublesome thoughts, goals, information, and actions, with healthier ones to end or avoid symptoms.

It may be helpful to metaphorically view therapy like a software update.

Not everything has to be changed, only that which is problematic.

People's personal software that's crucial in decoding and treating persistent depressive disorder has a very familiar name.

Personality

Dysthymic depression has many similar and interchangeable aliases, such as dysthymia, and persistent depressive disorder. However, it's the *neurotic* depression designation that's perhaps the most informative clue in this mystery.

Unfortunately, the *neurotic* word is often misunderstood. People may regard it – and use it inappropriately – as an insulting and disparaging term. But, it isn't…or at least it doesn't have to be.

Rather, think about the following to distill a more accurate definition.
1. Communication and consensus lead to *general* principles that apply to everyone in a society.
2. Within a society there are *subsets* of rules that concern *divisions* within it. For instance, various religious or political groups have more specific tenets that extend, provide unique conventions, or directly oppose the common canons (e.g., being pro-life in a pro-choice society).

Okay, but consider *the most fundamental rules* a person can have.

As you know, it's an individual's software, the distinct composition of goals, perceptions, guidelines, data, morals, etc. that creates each person's unique sense of reality and how to think, act, and feel in a social and physical world.

This software has a most recognizable name…Personality.

So, consider what this has to do with defining the term "neurotic".

Neuroses can be equated to personality disorders.[197]

There are many forms of depression; neurotic depression is merely one.

Being able to see how personality is related to symptom development allows you to differentiate it from hardware-based depression.

Indeed, the term neurotic depression draws your attention to how personality factors can shape symptom timing, severity, presentation, frequency, and duration. Psychologists assess and treat people's personality characteristics that are related to the development and persistence of dysthymic disorder.

Fair enough, but there's an inherent challenge.

As the word "personality" represents the forest rather than the specific trees of goals, rules of thought, action, emotion, etc. that take root and flourish in it, you soon connect a conclusion and requirement perhaps as daunting as piecing together a surrealistic jigsaw puzzle.

Personality is a vague and generic term.

To decrypt dysthymic disorder, you must know descriptive data and details.

Life Style Convictions

To more fully explain personality, perhaps it's best to return to the game analogy from earlier.

Although games can exhilarate and entertain, consider why individual strategies are the essential characteristic of gaming.

To start, consider how all participants must obey the general rules of play. Yet, within those broad regulations, *players devise specific strategies and use particular tactics they believe will enable goal achievement*. For example, people may choose to play close to the net in tennis, move a knight first in chess, or attempt long passes on the football field.

In their pursuit of besting the competition, participants must formulate plans of action they believe will allow them to win. The interactions of various strategies make for dynamic gameplay. After all, certain plans and attempts are used against others, with each participant or side changing strategies and tactics in response to, or anticipation of, other players' actions. The strategies and tactics are employed with the belief (or sometimes just a strong hope) that they'll enable success.

Individual strategies are perhaps the most significant factor as to whether players reach their objectives, and therefore you can perceive them as the defining feature of games as well as various other competitions.

Ultimately, what would a game be without them?

Now consider the parallel between games and human behavior.

Within a society, each person is obligated to obey the general rules. Yet, within those broad regulations, subdivisions within a culture have more specific stipulations (e.g., religious, political, family). But, most important, individuals select and adopt various ways in which to operate – usually within the common guidelines.

Similar to how there are individual strategies within a game, each person has a style of living, an attitudinal set toward life composed of various life style convictions that influence perception, thought, action, and emotion and instruct how to belong in the world. The life style convictions define how people move from an inferior position to one of competence or significance.[198]

This leads you to another conclusion (which should sound a bit familiar).

The life style convictions are the defining feature of each person and perhaps the most significant factors as to whether people achieve their objectives in life.

To avoid confusion, it's time to clarify the important, but similar sounding terms: *life style*, *style of living*, and *life style convictions*.

You can consider the life style convictions as a *modus vivendi* (one's general rules for living such as, "Be in control"), rather than a *modus operandi* (the exact way to be in control in a certain situation).[199] (Hint: *Think of each life style conviction as a metaphorical tree trunk; and its limbs as the various ways in which that conviction is specifically displayed*.)

Life style convictions constitute one's life style (aka style of living), which has four constituents:[200]

- Self-concept (people's perceptions of what they are and are not). For instance, "I'm a nice person who protects the underdog", "I have integrity and want to be right", or "I'm not aggressive."
- Self-ideal (people's belief of what they should be or do to be significant, competent, satisfied, satiated, content, or superior). For example, "I should go to nursing school so I can help others and earn a living", "I should be good (or better)", or "I ought to have a family."
- *Weltbild* (how individuals perceive everything external to themselves including what others expect of them. It's a "picture of the world"). For instance, "The world is a cold, cruel place where others expect me to cater to them (although I won't…or shouldn't)", "People are exploitive, lazy, and incompetent", "My parents want me to be a good sibling", "My boss wants me to be perfect in everything", or "Men are rude and women are moody."

- Ethical convictions (moral guidelines [which may or may not reflect society's] that stipulate fitting behavior such as, "Be kind to others").

Knowing this, think about how people's self-concept and self-ideal can be related to symptom development.

Those with a greater difference between their self-ideal ("How I should be") and their self-concept ("How I see myself") tend to be depressed, anxious, as well as insecure.[201]

They believe they're not living up to what they ought to be.

From their perspective, they're falling behind or failing at life. For instance, when people believe they should have more friends, be wealthier, have more sex, or live in a better location, they may experience depression. (Keep in mind that symptoms might not be the last puzzle piece to fall into place and make a complete picture of dysthymia.)

Now, imagine all the goals that individuals set in various areas of living and their rules for achieving them. *People have numerous convictions to get through life and make sense of it.*

In addition, think about how although some or many of your rules may match those of your friends, family members, or even those you don't know, given the number and complexity of factors, *each person has a unique mosaic comprised of countless convictions.*

For example,
- Your career choice.
- Your ideal living situation.
- Your dating style and history.
- How and when, or if, you pay bills.
- How to handle the breakup of a relationship.
- How emotionally close you are to your friends.
- How, or if, you make eye contact with strangers.
- What you do when served the wrong meal in a restaurant.
- How you view sex and who qualifies as a permissible partner.
- The ways you interact with an inebriated and aggressive person.

Honestly, is there anyone you know who lives exactly like you?

Of course not. In fact, if you were cloned as an infant, raised in the same circumstances, have the same experiences, etc., odds are you and your clone would have measurable differences because each of you has individual choice. Or, think about how you would make choices now that you didn't earlier in your life, say in schooling, financial matters, dating, or sex.

Life style convictions guide perception, thought, action, and emotion as well as how people understand themselves, others, and the world. They constitute the playbooks for living, and therefore are incredibly persuasive. For example, they influence your goals and how to achieve them, shape your ethics, your humor, what movies you like, what type of person you marry or if you marry, what kind of friends you have, and hobbies you enjoy.

Life style convictions comprise one's life style.

However, you may know the concept of life style by its alias.

In short, the term life style is interchangeable with the word "personality".[202]

Oh boy. This seems as needlessly redundant and potentially confusing as submarine sandwich synonyms (or the countless, colorful ways to say "drunk"). Alas, multiple terms for the same or similar concept is common in psychology, e.g., fundamental attribution error, correspondence bias, attribution effect. (Use whatever terms and metaphors you find helpful; there's many to choose from.)

Here's a summary to clarify the different names and concepts:
- One term for dysthymia is *neurotic* depression.
- *Neuroses* are equated with *personality* disorders.
- *Personality* is interchangeable with the terms *life style*, *style of living*, and the concept of *personal software*.
- Like how lines of computer code comprise software that creates the reality of the device and influences hardware, *personality* is like an individual's software comprised of *life style convictions*.

- Or, you may view personality as a forest, life style convictions as tree trunks, and the limbs as the ways those convictions specifically present.
- *Personality* represents each person's distinct perception of reality, influences body hardware, guides thought, action, and emotion, including how to address the life tasks.
- Life style convictions can be the biggest factor as to whether people achieve their goals.
- Each person's set of perceptions and rules for living generally exists within common sense, *yet is more uniquely identifying than fingerprints.*
- To know someone's personality (set of life style convictions) is to *genuinely and accurately understand the person.*

Now, recall from earlier that the literal translation of the word "psychology" is "study of the soul". Intriguing, yes, but consider the following to figure out that that means.
- When people say they found their "soulmate", they're declaring that they found someone whose characteristics (life style convictions and goals) are a perfect match for their own.
- "Soul searching" means a thorough examination of one's morality, motives, emotions, actions, perceptions, beliefs, objectives, etc.
- Individuals often describe someone's "soul" – at least in part – as the unique essence of a person's character (cognitively, emotionally, etc.).

Forgoing the metaphysical aspect, the word "soul" roughly describes the defining constituents of *personality*. In part, psychology is the study of how personality is used to address and achieve competency in the life tasks.

Now consider how this is relevant to persistent depressive disorder.

Every individual holds life style convictions believed to be true and acted upon accordingly. The paths people choose are selected precisely because they're believed to be the most efficient and effective means of goal achievement, *even if they're disastrous*.[203] In other words, people can have life style convictions they believe are realistic and helpful, and act in agreement with them even when they're related to symptom development.

When life style convictions don't fit environmental requirements, people can experience symptoms; which are, in part, a feedback mechanism to let them know that their way of thinking needs to change for them to be healthier.

Therapy assesses and treats people's personality and related goals.

This is logical, but there's a catch.

Life Style Convictions Can Be Strong and Inflexible

It's easy to say that life style convictions guide perception, thought, emotion, action, and goals; directing each person's sense of reality. Yet, these words can be a bit broad and blurry – as if trying to describe the Mona Lisa while looking through waxed paper – and thus insufficient to convey their influence and stubborn persistence.

Perhaps an analogy may be helpful to better understand their meaning as well as recognize and comprehend their strength and resilience.

Imagine obeying a simple rule while you're driving: Never deviate more than 1 MPH from the speed limit. Likely, you'd be so focused on rigidly following that rule that you'd probably end up driving into something which you shouldn't.

Flexibility of one's rules and the ability to adapt is healthy.

Just think about those whose life style convictions lead them to obey superstitions and compulsions. For example, they may knock on wood, cross their fingers, wash their hands excessively, repeat certain words, repetitively check door locks, verify that the oven is off, make sure that the TV volume is always set on a certain number, or avoid touching door handles.

People with superstitious behavior and compulsions can feel depressed, anxious, or angry that they're seemingly powerless over their rituals and urges. To them, they *must* do those actions, often deploy them reflexively, and find it as difficult to stop as gravity. Consequently, their compulsions or superstitions appear control them with the relentlessness of an overbearing, perfectionistic boss.

The force and frequency by which people adhere to their life style convictions is a testament to their dominance and hint at how they relate to symptoms.

Okay, but examine why people engage in these compulsive behaviors.

If you say those actions are random and rudderless, you'd be mistaken.

While such compelled acts appear to be illogical, costly, and painfully counterproductive, most people are able to articulate why they do them.

Think goal orientation.

They may fear that if they *didn't* engage in their compulsions, something dreadful will happen; for instance, their house may burn down, a loved one may die, they won't get a job interview, etc.

So, by obeying their compulsions and superstitions they strongly (yet, incorrectly) believe they have control of their environment – in some cases, over life and death. Simply, they're a means of attaining their desired outcomes according to selected rules and perceptions.

These examples show life style convictions' power and resilience as well as validate the universality and importance of human goal orientation.

Yet, some people might say, "These are intense and obvious thoughts and actions – and given that most people don't have compulsions and superstitions, the conclusions don't apply!"

Yes, these are extreme examples that glaringly go against the norm. But, such clearly observable demonstrations make it easier to show you how strongly compliant people can be to their rules of thoughts, emotions, and behavior. The degree of intensity is different, but the psychological mechanics are analogous.

People can have virtually automatic, unhesitating, and unquestioning obedience to their life style convictions.

If you remain in doubt about the concept's legitimacy or relevance, perhaps more personal examples may help illustrate the process.

Have you ever experienced any of the following?

- You're in a foreign land and notice people "do things the wrong way."
- You're in a discussion with someone of a different political or religious affiliation and think that the other person is exasperatingly naive, unenlightened, or misinformed.

- You can't relax on vacation because your routine is "out of sync".
- You're annoyed when a family member sits at your dinner table chair.
- You feel odd seeing a person wear something you never would.

Perhaps you've had such experiences because your life style convictions and limbs of related actions were changed, challenged, or violated in some way.

The life style convictions are so intertwined with perception that it's rather difficult to see their degree of influence.

Most individuals are unaware of the dominance of their own convictions when they're surrounded by others who think and act similarly, as that way of living is considered "normal" and goes unopposed.

Yet, the distinctions become obvious when others hold dissimilar life style convictions. For instance, you may have been at family gatherings with relatives who vocalize contrasting education, relationship, or parenting views and practices, and you felt irritated, frustrated, or angry and were tempted to say something in defiant opposition, or you just left a bit earlier than planned.

People can become increasingly cognizant of how fiercely and stealthily they hold their own convictions after interfacing with contrasting ones.

Knowing that life style convictions shape perception, imagine an individual has the belief that people are dangerous and self-serving and should be avoided.

Consider how this conviction can self-perpetuate and strengthen over time.

Sadly, that person is far less likely to be vulnerable and socialize. This retreat forbids interaction with peaceful, generous, caring others that could invalidate that unwelcoming and incorrect conviction, thereby fortify and maintain it.

Here are some more examples:

- Those who see a woman whose car is off the road and immediately think "Women drivers are incompetent", but when they see a man whose car is off of the road and default to "The poor guy's car has broken down."
- Those who perceive when people of a specific race or career are found guilty of a crime, but not notice when others are arrested or convicted.
- Individuals who believe that their furniture is superior because they bought it from an expensive, trendy store, when other, less pricey places have the same or better offerings.
- Football fans who are certain that a referee is of poor judgment and/or vision when a player on their favorite team is fouled, but perceive the referee as acting justly and perceptively when a player on the opposing team is identified and penalized for violating the game rules.

Given that life style convictions influence perception, thought, and action, they can silently and powerfully reinforce themselves by defining what's valued, devalued, and ignored.

Think about how life style convictions and connected branches of action can be powerful and resistant to change. For instance, consider how perfectionism allows people to get and keep a job, wanting predictability permits proper vacation planning, or being good avoids conflict.

When life style convictions enable goal attainment, the desired outcome reinforces them.

Now consider those imperfect convictions and rules of thinking and acting which are only able to guide people to their desired outcomes occasionally. You may guess that convictions as limited as a T. Rex playing Simon Says and would eventually go extinct because they don't always allow goal achievement.

But that isn't the case.

Life style convictions which allow goal achievement only some of the time are likely to maintain behavior (intermittent reinforcement).[204]

Also, similar to superstitions, people can misperceive the utility of their life style convictions and attribute goal achievement to them incorrectly.

Odds are, you can sing the alphabet song with the same cadence as when you learned it during childhood, tie your shoes or ride a bicycle nearly instinctively, or immediately call up dozens of actions and rules you acquired early on.

Early learned convictions and their associated actions can be well-rooted.

Now, recall how groups, societies, and countries may get others to adopt their "common sense" by shunning, imprisonment, and the lethal force of war.

With that in mind, think about how some parents *demand* that their children think and act like they do (e.g., have the same profession, be neat, follow particular sports [and perhaps the same team], dress in a certain way, maintain traditions, have the same religious or political beliefs), and who *threaten* their children in various ways (e.g., yelling, avoiding them, taking them out of the will, criticizing them in front of others) for violating those life style convictions.

Some parents say, "If you don't behave (abide by particular convictions and act in the manner I want you to) then I'll take away your allowance (not bring you to the movies, ban your friends from coming over, return those things I bought you, or threaten physical harm, such as spanking.).".

Threats for non-compliance to life style convictions are common and can occur throughout the life span.

For instance, some adults threaten their spouse with divorce and "never being able to see the kids again" or deny sex until some behavior changes.

Note that life style convictions can be contrasted with behavioral rules that are easily changed. For instance, as a rule you may take a certain road home from work, but by individual choice – or detour sign command – you readily alter your behavior with barely a whisper of distress. *As behavioral rules are flexible, they aren't related to symptom development*. Therefore, it's important to differentiate them from life style convictions.

When individuals rigidly and stubbornly clutch their life style convictions as if they were life-saving cliffside vines preventing a fall into a rocky abyss, it can be excruciatingly difficult for them to adapt to situational requirements and achieve healthy objectives in the life tasks, and it's more likely that they'll develop symptoms.

Dogmatic tenacity to one's convictions is a primary and defining characteristic of neurotic depression.[205]

When people's personal legislation vetoes goal-attaining action, they may erode their self-esteem, feel as powerless as prisoners, confined from peaceably and vulnerably interacting with others, as well as dogged by other punishing sensations and conclusions.

As if that weren't difficult enough, think about what it is about people's personalities that can magnify the probability of symptom development.

Subjective Convictions

People are social beings, interdependent on each other throughout the life span. Communication cultivates common sense which enables tremendous cooperation and successes when people share rules and perceptions (e.g., all pilots and air traffic controllers must speak in English or invite disaster).

People generally agree on, and act in accord with, common knowledge.

But there's an inherent difficulty when diverse individuals coexist.

Think about individual choice and call to mind those you find irritating and infuriating. Visualize and echo their exasperating characteristics and statements to identify the major factor associated with such friction.

Often, it's nothing more than *personality clashes* – the distressful differences when others think and act differently than what you expect or prefer. Frustration, anger, sadness, confusion, abrasion, division, and other evidence of grating incompatibilities can arise rapidly and, sometimes, quite painfully. This you know, but it hints at something that can be exceptionally aggravating.

Just consider how you feel when family members or valued others misperceived you, your thoughts, feelings, or intentions.

While you hold a certain self-image you believe is true, when others see you differently than you see yourself, your self-perception and sense of worth becomes one view of many, perhaps discounted and unnervingly subject to dissimilar interpretations and evaluations. Different perceptions and ways of living can be upsetting. ("Hell is — other people!", as Sartre once wrote.[206])

Such conflicts, differences in life style convictions, and diverse judgments are most apparent in gray, subjectively-judged matters of right or wrong, good or bad; for example, culture, literature, architecture, music, morals, philosophy, parenting, religion, politics, art, or movies. Perhaps you've argued with friends and loved ones over their (ahem) poor taste in music, movies, clothing, etc.

Through social influence and choice people nurture a unique set of life style convictions and the related style of living which have inherent biases. This can breed disagreement in subjective topics that may not have a clear answer.

Expanding a metaphor: Equate personality to a forest, life style convictions to individual tree trunks in a forest, and the limbs of a tree to the specific ways in which a conviction presents. Most trees in the forest are in good shape and function well, and all or nearly every branch on each tree is thriving happily.

Now consider where symptoms fit in.

Imagine that one conviction is to be in control. Most branches on that tree are appropriately useful (e.g., control spending, eating, schedule). Yet, one or more branches may be problematic (e.g., trying to control adult children or friends). Not every limb on a tree leads to a symptom, most are healthy and function well, but symptoms sprout from a life style conviction like limbs from the trunk of a tree. Therapy helps people identify and prune thorny branches.

The variability among people's contrasting personalities can end in distress, confusion, frustration, annoyance, conflict, heartbreak, etc. Sometimes, conflicting personalities are like different forests competing for the same land; or a couple who argue over different parenting styles.

Personality differences appear to be inherently and inescapably problematic and counterproductive, but think about when that isn't the case.

Different Convictions

Pause to consider the wonderful variety of plants and animals around you. In their various colors, actions, constructions, and contributions, the medley of flora and fauna is impressive, beneficial, and enjoyable. You can observe the beauty of birds in flight, taste the array of delicious fruits and vegetables, sit in the calm shade under the canopy of giant oak trees, behold an elephant's impressive stature, or note flowers' scented melodies.

Such a colorful palette of gratifying, life-affirming experiences breaks the gray monotony of similarity and routine. There are reasons for such variation.

It's because of a combination of genetic mutation and natural selection.[207]

Essentially, mutation is a change in the structure of a gene.

Sometimes, this can have a negative effect that lessens the ability to survive and/or thrive (e.g., blindness, lameness, inability to reproduce). Consequently, the gene is more likely to be selected out of the gene pool.

Other times, there can be no effect or a neutral effect, which doesn't impact – one way or another – the ability to survive and/or thrive (e.g., whether you have free or attached earlobes). These genes can quietly hitchhike on the genome and be carried on from generation to generation.

However, positive mutations can increase the ability to survive and/or reproduce.[208] For example, if one plant develops the ability to deter predators or significantly increase its fertility, it becomes more popular than its same-species siblings. Or, if one plant is better able to attract bees which aid pollination, or a tree produces desirable fruit which animals eat and later spread the consumed seeds far more widely than the tree ever could (usually with some nutrient-rich natural fertilizer).

But there's one more thing to think about.

Environmental changes occur (e.g., variations in available food, water supply, shade, temperature, number and types of predators). With that in mind, you can see how a lack of mutation can prohibit adaptation to the surroundings, which may decrease the likelihood of survival and/or propagation.

Consider how this relates to psychological symptoms.

Next, although mutations can be positive for certain plants or animals, they don't always have a useful outcome for people. Sometimes genetic deviations can also lead to most unpleasant things for humans, such as poison ivy, mosquitos, or lethal viral infections.

This metaphorically suggests something about personality differences that relate to persistent depression.

Biodiversity Clues

First, plants and animals may have genetic changes that beneficially allow adaptation to environmental changes. Likewise, each person needs to adapt *psychologically* and behaviorally to environmental changes. Consider how humanity moved from a hunter-gatherer culture, to an agricultural environment, to an industrial setting, to an information and services-based society, and so on. Or, think about how some people must adapt from being single to being married with children to being divorced or widowed.

It's most desirable to avoid arrested psychological development – cognitive rigidity (say, being dogmatic to life style convictions). *When individuals cannot psychologically adapt, they imperil goal achievement and may become angry, anxious, depressed, etc.* In some cases, the outcome can be lethal (e.g., depressive spiral to suicide, inadvertent death from overconsumption).

To survive and thrive in evolving and fluctuating environments, people must adjust their convictions and behavioral rules to adapt psychologically.

Second, think about how deviations in genes can lead to distinctions within a species (e.g., different types of apple trees or butterflies).

Consider how this relates to life style convictions.

A person who thinks in a way that differs from common sense can create a mild or noticeable distinction (e.g., new words [see Shakespeare], culture, dress, political or religious ideology and practice).

Third, consider how people can select a certain type of apple tree and make an orchard of them. The deliberate picking and reproducing a particular species is known as artificial selection, which can increase a species' frequency.

Analogously, people can choose existing convictions, thoughts, and actions, and prompt others to replicate them, thus make them more common.

Fourth, think about how different variants of apple trees have established over time. Although there was a common genetic ancestor, due to mutation as well as natural and artificial selection, noticeable distinctions grew.

With that in mind, you can see how divergence among related, but different convictions and behavior, can become entrenched.

Just consider how siblings who grew up in the same household with identical holiday traditions may slightly alter those customs, favoring and replicating certain ones, while minimizing or discarding undesired other ones, which lead to slight, but recognizable alterations from a common past.

Modest changes in shared convictions can be repeated and lead to noticeable differences that take on a distinct life of their own. (For instance, did you know that there are more than 33,000 denominations of Christianity?[209])

Slight variations in convictions, perceptions, thoughts, and behaviors can lead to perceptible differences within groups or cultures.

Fifth, recollect how a positive genetic mutation is a deviation from the norm. Although it's a numerical minority in the population, in time it can become the forerunner of change and eventually the prevailing, defining template.

Similarly, a way of thinking can start off as a deviation from the custom or standard practice (an ideological outlier, a statistical minority), but become widely adopted and part of the common sense (e.g., safe sex, exercise).

Deviations of convictions and action can grow from a minority to a majority.

Sixth, genetic divergence can be so different from the original, drastically dissimilar species arise (e.g., apple, pear, cherry, and orange trees).

Consider what this suggest about people's convictions and behaviors.

Differences that become part of the common sense can lead to vastly dissimilar common sense among cultures or nations.

Seventh, recall how genetic mutations can lead to harmful manifestations.

Analogously, divergences of thinking, say an inadequate perception or problematic rule about personal hygiene, eating, exercise, saving money, or socializing can lead to obesity, low self-esteem, loneliness, self-harm, being unpleasant or mean, promiscuity, failure, prone to being exploited, or a fatal outcome (e.g., premature heart failure, lethal infection).

Life style convictions can be counterproductive, unhealthy, and deadly.

Eighth, think about what the following reveal.
- Fish fight for hiding spots from predators.
- Sperm jockey with one another to reach an egg.
- A litter of pups scuffle over contact with their mother.
- Male deer fiercely challenge each other for access to females.
- Plants that consume nitrogen into their roots and starve nearby rivals.
- Plants compete (e.g., in scent and color variations) to attract pollinators.

There's unrelenting competition for limited resources within and among species, which can jeopardize the ability to survive and thrive.

Knowing that, examine the list below for revealing clues about humanity.
- Political debates and elections.
- Sibling rivalry for family attention and pride.
- Different corporate strategies among businesses.
- Religions that advocate different gods, rules, and rituals.
- Various and incompatible economic philosophies and policies.

People have different convictions and ways of acting and, by extension, distinct common sense, in ceaseless competition over limited resources (e.g., money, people) which can imperil their ability to survive and thrive.

People can have differences that can be annoying, aggravating, symptom producing, or lethal. Such dissimilarities don't always lead to disagreement, frustration, pain, sadness, and imperil people's ability to survive and thrive.

Just consider plants and animals *within a species* that assist each other (e.g., grooming, hunting, alerting of nearby predators). Even *different species* help each other. For example, bears and magpies help each other hunt, animals and plants can benefit each other as do humans and other animals, say when people feed and shelter dogs, and dogs can provide companionship, assist hunting, and be service dogs. *Genetic deviations allow amazing and delightful plants, animals, and experiences, including mutually-beneficial cooperation.*

Simply, those imperfections in genetic replication called mutations enable harmony and growth that may not occur otherwise. Now consider how this genetic analogy applies to psychology.

People have different and incompatible life style convictions, etc., which can be divisive, counterproductive, and infuriating. Yet, consider how as each person has a unique view and distinct strengths, no one can know everything or take every perspective. Indeed, the number of understandings of life is equal to the number people with each understanding having inherent mistakes, a margin of error, *as there's no one right way or totally wrong way of living.*[210]

In other words, no one thinks exactly as you do. Moreover, no human being can be omniscient and infallible – even when the person truly believes that the held perceptions and rules for living are impeccably correct and realistic. There's not a perfection of thought, only various imperfections.

Every person has knowledge gaps and ranges in accuracy of thought.

Tellingly, the human solution to this challenge isn't terribly different from the genetic one.

Consider how varied life style convictions lead people to develop diverse assets and goals that allow them to fulfill various roles and work as a coherent unit toward a goal, much like any team (though some team members may procrastinate, aggravate, rebel, or don't contribute significantly or at all, ugh). Just recall a time when you couldn't solve a problem or puzzle, but another person with a different point of view provided a solution or a novel approach.

On a larger scale, when people communicate their various perspectives, they can negate each other's errors and reach a consensus. This allows for correction of not only the individual's information, but also the refinement of the group's shared knowledge. Common knowledge is usually, *but not always* a better, more realistic interpretation of reality that enables harmony and goal achievement (e.g., popular misconceptions, urban myths, unhealthy folklore).

So, just like having access to different types of food that provide all the necessary nutrients for people to be healthy and grow, a pool of common – but variable – knowledge allows an encyclopedic library from which to solve life's problems and enjoy one another's contributions.

Similar to genetic variation, there are benefits when people have different convictions. Sometimes the class clown becomes the creative artist or comedian who continues to think and act differently to inspire, move, charm, or cheer up the nine-to-fivers who robotically replicate what's established.

An array of life style convictions (personalities) is a tremendous advantage for the human race, as differences in thought and action allow for various roles and solutions to life's problems and propel humanity forward.

Yet, consider how if everyone thought and acted the same way, it would be impossible to adapt and come up with new solutions. This would threaten humanity's ability to attain its two biggest goals, to survive and thrive.

As frustrating as different personalities may be, they enable better approximation of reality, increase goal achievement, among other benefits.

There can be perfection in imperfection…*but not always.*

Recall how negative genetic mutations and a lack of mutations may prohibit development and imperil survival and thriving, especially in a fluctuating environment.

This hints something about psychological symptoms.

When a person's convictions are unhealthy deviations from common sense, or inflexible and unable to adapt to environmental requirements thus hinder or prohibit goal attainment, this can relate to developing neurotic depression.

Thankfully, there are many ways out of the symptom maze.

One example is that non-symptomatic persons provide convictions that *aren't* associated with symptom development. This can be used to assess where dysthymic persons' convictions and objectives problematically differ.

With that in mind, the following becomes logical and apparent when discussing the discipline whose "study of the soul" focuses on personality factors, social connection, goal orientation, and the action or inaction to achieve countless and various desires.

Psychology examines a person's movement and objectives in comparison to others.[211]

There's a name for people's set of perceptions, values, rules, and goals they believe will enable competency, but also can be related to developing depression.

Fictional Finalism

To survive and thrive, everyone must address the life tasks. Yet, each person starts life in an inferior position. So, everybody must select goals as well as the perceptions and paths to achieve them, with the belief they'll be competent after reaching their chosen ambitions.

This guiding image is known as *fictional finalism*.[212]

It represents many things, such as...

- People's long-term goals that enable them to be competent in life.
- Accomplishments believed to grant a sense of worth and belonging.
- A depiction of how their life will be after achieving those objectives.
- A representation of how they believe the world should be.
- Something that influences people's perceptions, values, and actions.
- A general conception which people may not be fully conscious or able to verbalize completely.
- An impossible-to-achieve ideal.

For instance, someone may believe that being a good person, working hard, getting married, having a family, maintaining friendships, vacationing regularly, and retiring early, is a template for being competent in life.

The set of goals and beliefs in fictional finalism can be somewhat vague.

In other words, there aren't strict definitions of how to be good, what career to have, who to marry, how big of a family to create, what type of friendships or how many friends, or where to vacation and how often.

Think of *finalism* like a light on a distant harbor that silently pilots people toward a somewhat foggy destination and allows them to avoid an unnerving sense of being rudderless in life, without goals or guidelines.

Given that people are goal oriented, this is unsurprising. Think of how children or young adults want to be a medical doctor, and use that to guide them in their training, self-perception, what books they read, what TV shows they find interesting, etc. – all without specifically defining what type of medical doctor to be, how or where to practice, and so on.

Okay, but consider why it's a fiction. After all, people live in reality, experience tangible failure, have genuine friends, get an actual job, etc.

To start, you can think of the initial state (being inferior and unable to address the life tasks) *as a minus situation* and the goal state (competency or superiority) *as a plus situation*.[213]

Knowing there's individual choice, social influences (e.g., family, society), different environmental advantages and limitations, people have the desire to overcome actual and perceived minus conditions in various creative ways.[214]

Each person crafts a unique narrative of how to perceive and get through life. Given that this guiding goal is a creation, based on subjective perceptions, rules, etc., and may not necessarily be realistic or true, it's a *fiction*.

Fictional finalism can occur within the parameters of commonly accepted means and do not necessarily indicate psychopathology.

People may be frustrated or depressed, say when they perceive themselves as not on track or falling short of living up to their fictional finalism – or when others and the world do not turn out as predicted or preferred.

As people may not be fully conscious of their fictional finalism, therapy can help them understand how it's related to symptom development. In fact, there's a significant, descriptive concept which is related to fictional finalism that individuals use to assess competency.

"Real" Man and "Real" Woman

The subjective conceptions of a "Real" man or a "Real" woman represent each person's unique definition of ideal characteristics that enable goal achievement (competency).[215]

You may think of it as a concept that's more focused than fictional finalism.

For instance, someone may define a "Real" man as someone who works hard, gets married, has many children, lives in a lavish mansion, and saves $5 million for retirement. Someone else can define a "Real" man as someone who has many sexual partners throughout life, spends money recklessly, drifts from job to job, enjoys various drugs and drink, avoids paying taxes, rents instead of owns, and is never tethered to any person, place, or thing.

There isn't a universal standard, and due to the sheer number of variables involved, *there are countless definitions of a "Real" man and "Real" woman.*

Next, consider those who define a "Real" man or a "Real" woman as a drug dealer, thief, sex worker, or fraud.

The concept of a "Real" man or a "Real" woman can include unhealthy and socially harmful characteristics and goals.

Given that the characteristics that define a "Real" man or a "Real" woman are seen as an ideal, these conceptions may be used as not only as something to strive for, *but also as a yardstick of worth of self and others that leads to an expression of admiration or condemnation, acceptance or rejection.*

Falling short of being a "Real" man or a "Real" woman may have particular implications for depression. For example, the greater the difference between the self-image (how people see themselves) and the self-ideal (what individuals believe they should be) leads to increased, and perhaps significant, inferiority feelings and a sense of being overwhelmed, ineffective, unintelligent, frustrated, or defective in some way.

Sometimes, parents may be depressed when they perceive their children as not living up to the parents' definition of a "Real" man or a "Real" woman (e.g., living at home at age 28, unmarried, underemployed, and childless).

Likewise, some individuals may be depressed when their siblings, friends, or parents fall short of the defined characteristics of a "Real" man or "Real" woman. Indeed, people can be dismayed when friends and family aren't as supportive or attentive as expected (e.g., they don't want to help watch the grandchildren, they don't go to a birthday party or wedding – or leave early, or they do not offer financial assistance during a difficult time).

On a larger scale, political or religious mindsets may define the concept of a "Real" man and a "Real" woman (e.g., do people act in accord with preached or propagated characteristics born from one point of view).

Even though "Real" man and "Real" woman definitions are subjective, people may act as if they were unchangeable and permanent, which may compound a depressing sense of hopelessness.

The subjective creations of fictional finalism, "Real" man and "Real" woman, selected perceptions, life style convictions, and desired objectives can be related to developing dysthymic depression.

But keep in mind, these chosen, sometimes problematic, factors help people do something vital.

Adapt, Survive, and Thrive

Recollect that the combination of genetic mutation, natural selection, and evolution, enables genes that have a neutral or positive influence on a species' ability to live, prosper, and endure, while those that diminish the probability of surviving or thriving decrease in quantity or fail to remain in the gene pool.

Animals have DNA-based behavior that greatly aid their survival and thriving (e.g., swimming, spinning webs, building nests, migration). Genetic changes have enabled a gradual adaptation to their surroundings, acquiring various characteristics and abilities that help them survive and thrive.

People have a number of evolutionary advantages as well, but they don't have such a broad array of innately-inscribed behavioral solutions as other species. While other animals can learn various and nuanced things (as most dog owners can testify), there's a hardware solution that seems to give them an extraordinary edge over human beings who do not have such inborn instincts.

Yet, imagine how people's seeming shortage of innate solutions to the inescapable life tasks can be advantageous.

To start, think about what's likely to occur if your phone or computer left the factory with unalterable software.

Those devices couldn't adapt to various environmental threats and demands (e.g., a virus), or have their performance enhanced (e.g., run faster, increased functionality of an app or operating system), or correct the computer code that hampers performance (e.g., software "bugs"). The incorrect data and inflexible rules and goals of operation quickly make those devices obsolete dinosaurs and soon discarded, as their inability to evolve can create a mass extinction.

But there's a solution.

The rapid and frequent distribution of software updates.

Computers and phones can swiftly adapt how they operate. This enables them to be immunized against viruses, squash bugs, and improve functionality, which increases their useable lifespan.

With this in mind, you may be able to see the advantage of human beings' relative lack of inborn solutions to the life tasks.

For humanity to have the best chance of surviving and thriving, people must be able to live in as many environments around the globe as possible and adapt to various environmental threats and demands or imperil their existence.

Recall that there are a number of human genetic changes (e.g., lung capacity, skin color) that have enabled adaptation to environments.

However, there's a difficulty with this hardware solution.

Genetic mutation is random, and natural selection and adaptation can be glacially slow. Therefore, a hardware solution to environmental demands can be dangerously sluggish, inefficient, and insufficient. Yet, human beings are more plentiful and widespread than McDonald's and Starbucks.

They aren't dependent on a stable or perilously-narrow setting. They live and multiply in variable conditions and diverse environments, such as deserts, jungles, tundras, cities, swamps, mountains, and countryside.

And recognize what has allowed people to adapt, survive, and thrive in various environments around the globe and competently fulfill the life tasks.

Relatively rapid and frequent distribution of software updates.

People's ability to accumulate, incorporate, and transfer knowledge that's relevant to their surroundings (e.g., cultural customs, such as planting practices and harvest rituals) has multiplied the ability to adapt, survive, and thrive. Sociocultural evolution occurs at a much faster rate than genetic evolution.[216]

Although the relative lack of DNA-based instincts initially appears to be an imperiling weakness, a software solution – the ability to create, maintain, and transfer new and appropriate life style convictions – has allowed humanity to competently address the life tasks across time and geography.

For people to adapt, survive, and thrive they must be competent in their physical and social environments. Yet, as they're born in an inferior position, they must – *with each person in every generation* – adopt goals and acquire the ways of thinking and skills they believe will allow them to address the life tasks.

Think about how in natural selection *environmental* factors (e.g., physical setting, predators) influence what genes remain. However, with artificial selection (aka selective breeding), *people choose* what gets replicated and multiplied in the gene pool. For example, there has been deliberate selective breeding of corn, rice, bananas, wheat, dogs, and cows.

In fact, people do this with other human beings.

Individuals are routinely attracted to others with particular physical characteristics – such as eye color, skin color, hair color or texture, or height – because they want their future children to inherit those qualities.

This relates to life style convictions and neurotic depression.

The goals, ways of thinking, and the paths people choose depend on a number of factors: impressions and sensations from the environment as well as their bodies, and the interpretations of that data, as well as their education (others' direction and guidance).[217] For example, the media, environmental conditions, family, friends, etc. can be influential factors when people pick their objectives and how to achieve them.

Yet, think about the impact of individual choice. For instance, are you exactly like your parents, or are there a few things that you vowed a long time ago that you'd never replicate?

While a restaurant provides the menu, ultimately you choose your meal from it or go elsewhere to satisfy your appetite.

Although social factors are influential, individual choice determines what life style convictions take root and grow.

Analogies can effectively communicate concepts. With that in mind, note that the use of the terms adaptation, mutation, evolution, natural and artificial selection was deliberate…and perhaps a bit more relevant than you may have imagined.

The view of human life as a progression from an inferior position (e.g., unknowing of the world, smaller, weaker, less verbal, and with a complete reliance on others for survival) to a goal state of competence in the life tasks, by social interaction and interdependence, enables people to adapt to their environment, thereby survive and thrive, is Darwinian in character.[218] Indeed, using evolution as the foundation for understanding, you can see that people strive toward their goals physically and psychologically.[219]

Thoughts and actions that enable goal achievement survive and thrive, whereas those that are unproductive are employed less frequently or go extinct.

This doesn't mean people's accumulated life style convictions and related objectives and actions are always healthy. Just envision a child who regularly claims inability or unfairness, cries, becomes lethargic, and is pampered in various ways (e.g., given undue attention and preferential treatment, permission to avoid challenging situations, others take over the child's responsibilities, such as homework or chores). The child may maintain that unhealthy way of thinking, emoting, and acting. And think about what might happen if this situation occurs over decades, with various people enabling and reinforcing the behavior (e.g., siblings, spouse, friends).

The life style convictions may become entrenched, and the related symptoms can arise with increased quickness, strength, and frequency.

Indeed, many adults continue to live with their parents who pamper and protect them. The environment may enable the perpetuation of individually-chosen unhealthy convictions.

Obversely, imagine the child who works hard, whose parents and teachers encourage perseverance through difficulties, and whose convictions and actions are reinforced (e.g., an empowering sense of accomplishment and ability, parental praise, social acceptance and elevated status, money, receiving a desired gift or some benefits others do not).

Here too, you can see how life style convictions, associated actions and emotions, as well as goals can be maintained and strengthened. (Tellingly, sometimes the industrious, courageous kid is the sibling of the one who retreats from challenges – this also shows how individual choice is a crucial, decision-tilting factor.)

While such socially-accepted and encouraged convictions and actions appear to make people immune to symptom development, that isn't always true.

Depression may arise when industrious, kind, thoughtful individuals have unrealistic expectations of others who do not act in the same manner and fail to reciprocate compassionate or dedicated actions.

Knowing life style convictions and goals are subjective creations used to adapt, survive, and thrive allow you to realize something relevant to persistent depression.

Revealing Choices

Resolving the mystery of dysthymic depression requires that you appreciate different mindsets and approaches, as each can provide valuable clues.

There can be physical, physiological, biological, chemical, and genetic causes of depression. Psychiatrists perceptively identify innate issues as well as capably and compassionately address bodily hardware that trigger symptoms which stop people from taking healthier action.

But there's a risk to this well-known practice.

First, consider how people must fulfill various time-consuming obligations (e.g., parenting, working, socializing, chores). Consequently, they usually lack the time to thoroughly investigate and have extensive knowledge of various things (e.g., how to do various types of home repair, file their taxes, make a gourmet meal). So, they may sensibly rely on experts for guidance and help.

Second, given time constraints and insufficient training, it's understandable why people depend on quickly-accessible and easy-to-grasp information – especially that which many others believe and support.

This reveals one of the brain's problematic trapdoors.

When it comes to understanding complex things, people can show a strong tendency to rely on and trust familiar and cognitively-easy concepts.[220]

With this in mind, think about dysthymic persons whose friends and family (or popular media) endlessly emphasize and echo certain phrases and notions about what causes and perpetuates depression as well as regularly recommend related remedies. A concept that's been repeated, or primed, is more likely to seem true even if it's false. (See: urban legends, myths, and folk remedies.)

Also, consider how often people boil down complicated and nuanced situations to simple solutions and statements. Such inclinations are seen in bumper stickers, online postings, or t-shirts with a catchy motto, for example, that advocate a particular mindset and associated action.

Readily-generated or recalled concepts can be naively applied and recklessly overused.

Those without specialized training, knowledge, or practice may quickly and incuriously default to and believe popular perspectives and simple answers. Painting only with primary colors doesn't allow an accurate representation of reality. Some laypersons may oversimplify and overextend science-based perspectives and practices that experts constructively employ.

As an illustration, consider how biological determinism is the belief that people's genes or physiology or some other innate factor largely or completely control their mental and physical activity.

This mindset can lead to extreme and troublesome outcomes.

Just consider how people commonly attribute others' complicated actions and cognitions to their sex (e.g., "All women are scatterbrained, moody, gold-diggers", "All men are sex-hungry, self-centered, jerks"). Racism provides additional clear and large-scale evidence of such thinking and related dangers. These examples provide a glimpse of how often simple conceptions, such as hormones or genetics, are blamed for people's actions or lead to the use of blanket statements to explain complex thought patterns and behavior.

Think about how this tendency to default to a simplistic, familiar conception and related solutions may be most likely applied to depression.

It's when dysthymic individuals' statements and actions seemingly conflict. When speech and behavior appear opposed, it can misdirect people's attention. For example, those with persistent depressive disorder may state – sometimes quite passionately – that they want to date, socialize, or work, but their symptoms puzzlingly seem to sabotage their efforts, paralyzing them from acting in accord with those stated goals. Thankfully, such ostensibly incompatible factors can be deciphered and understood.

When people default to biological determinism without looking further, they may deprive themselves of perceiving a panorama of factors that provide a complete picture of dysthymic depression.

Differential diagnosis is the process of discriminating among various factors to identify those which account for people's symptoms. While psychiatrists largely focus on hardware issues that cause depression, psychologists diagnose and debug the software. But there's a notable challenge.

As psychology assesses and treats the intangible, nuanced, complicated and somewhat abstract, it can appear to some as impractical, uncomfortably vague, and unscientific. Therefore, some people dismiss it without investigation.

Yet, consider how psychology is like a Ferrari, the Sydney Opera House, or iPhone. All are a marriage of science and art.

Psychologists use valid scientific methods (e.g., standardized psychological assessments), are trained in identifying bias and errors in logic, and employ research-supported and refined interventions.

Psychology is a science.

Yet, psychological assessment and treatment are dissimilar to mechanical engineering, for instance, that's filled with tangible, easily measurable components. There's something relevant that's beyond the tactile.

Take a moment to consider that when people enjoy music, they usually don't dissect a lively song into notes, and perform an isolating, tone-deaf autopsy. Likewise, people tend to appreciate a painting as a single, complete image, and attend less to the individual brushstrokes.

There's a gestalt, the whole is more than the sum of its parts.

Analogously, psychological functioning and symptoms are extraordinarily complex. Like studying physics, the variables are often intangible and quiet, yet once detected and decoded can communicate complicated concepts and meanings. There must be an awareness and respect of subtleties as well as nuanced approaches.

To see the big picture, you need something in addition to tangible science.

Something a bit more sensitive and intuitive, that still has logic and cadence, but adds illuminating color to otherwise black-and-white notes.

Recall that psychologists are like detectives. They must notice the soft, gentle hints in speech, action, and emotion. For example, just consider what's spoken by a nervous knee bouncing in rhythm with one's tension level, one's rate of speech, word choice, posture, avoided questions, amount and direction of eye contact, or how a newlywed couple sit intensely hugging a nearby arm…of the couch, as they icily sit as far away as possible from each other.

If cognitions, emotions, and actions be like a marching band in which all musicians play the same song and cooperatively move forward in harmony, consider what this might reveal about conflict between dysthymic individuals' statements and actions.

Psychologists combine what they detect with research and must present that information in a way that's likely to resonate, be understood, and memorable.

While noticing subtle signs is useful, gentle articulation is also required.

It can be hard to accept certain observations and evidence. People can be prone to experience hurt feelings or develop anger, for instance, when given details that's counter to their perspectives or beliefs. They may be sensitive to that which reveals their role in symptom development or actions that led to failure or rejection.

For example, people want to be good parents and can be confused when their children don't speak to them. Identifying how they have alienated their children (e.g., by being too controlling, or angry) may be excruciating for them to hear.

In addition, some interpret clinical observations and advice as imperishable proof of inferiority. Psychotherapy requires awareness and respectful, delicate communication. Otherwise, suffering individuals may counterproductively end treatment as if escaping a theater showing a clever, riveting movie, but filled with an abrasive, noisy audience.

Psychology is also an art – a teachable, learnable art, but an art nonetheless.

To comprehend psychologically-based neurotic depression, it's necessary to investigate how people's chosen compilation of life style convictions and goals they use to adapt, survive, and thrive relate to their symptoms.

Psychology is a science and art that can be used to detect subtle evidence. The gathered information is employed to decipher and comprehend seemingly irreconcilable factors and diagnose the software factors that explain why symptoms exist and guide how to resolve them. This can decrease people's proneness to rely on overly simplistic concepts and familiar terms that risk misdirection and mistaken conclusions.

How people choose to answer the life tasks provides a distinct and revealing collection of clues about who they are. To comprehend and treat neurotic depression it's necessary to note that the goals people select and how they go about achieving them create uniquely identifying psychological fingerprints and simultaneously lessen the grip of biological determinism.

Notebook of Clues and Evidence, Part Two

Psychology is complex; filled with concepts and helpful practices. In your progressive decoding of persistent depression, you've encountered a mystery's worth of terms, ideas, and hints. So, reviewing clues and evidence may allow you to take a much-needed break from investigative discovery and connect the dots which coax a revealing picture of symptoms to emerge.

1) CHALLENGE: Human beings must determine how to survive and thrive despite their relative lack of DNA-inscribed instincts and actions.
2) SOLUTION: Common sense
    a) Communication and consensus cultivate common knowledge of generally accepted and encouraged rules and goals that evolve over time.
    b) Cooperation allows the creation of shared guidelines necessary for coherent interaction as well as individual and group goal achievement.
    c) Largely, people's rules and perceptions by which they live are subjective creations projected onto reality and acted upon as if they were true.
    d) The ability to create, maintain, and transfer new and appropriate information has allowed individuals, as well as humanity in general, to competently address the life tasks across time and geography.
    e) Social and institutional reinforcement and punishment to enforce compliance to common knowledge reveals the significance of similar perceptions of reality, thinking, acting, and goals.
    f) Common sense influences how people define mental wellness, as well as comprehend and treat psychological symptoms.
3) PROBLEM: Common sense can be inaccurate, counterproductive, or unhealthy.
4) PROBLEM: Common sense may not be as common as people expect or prefer.

5) CHALLENGE: Individuals must determine how to move from an initial state of being unable to sufficiently address the life tasks to competency. This is movement from a minus situation to a plus situation.
6) SOLUTION: Life style convictions
   a) Similar to how common sense guides a community, each person selects a unique set of life style convictions that influences thought, perception, action, emotion, and goal selection, and instructs how to think, act, and feel in a social and physical world to be competent.
   b) Self-determination in goal selection and the rules to achieve chosen objectives starts at a very early age.
   c) Given individual choice, social influences, and different environmental advantages and limitations, people have the desire to overcome actual and perceived minus conditions by various creative ways.
   d) While common sense provides readily-available rules and objectives, and social factors can be influential, it's individual choice that ultimately determines which life style convictions are chosen or created.
   e) People's life style convictions are perhaps the most significant factor as to whether they achieve their objectives.
   f) Reinforced life style convictions and goals take root and grow.
7) Life style convictions compose each person's personality (aka software, life style, or style of life).
   a) Likening personality to software, life style convictions are akin to lines of software code that define reality, shape perception, action, and goals.
   b) People's software can activate or suppress specific hardware, (e.g., areas of the brain, muscles, hormones, neurotransmitters) to reach a goal.
   c) People's hardware and software can work together.
8) BENEFIT: Diverse personalities enable a wide range of roles and solutions to life's challenges. They can rocket humanity forward by

allowing a better approximation of reality as well as different and novel solutions that guard against replicating narrow and unsound beliefs.
9) PROBLEM: Contrasting personalities can lead to confusion, frustration, annoyance, conflict, and heartbreak, among other undesirable outcomes.
10) PROBLEM: Life style convictions can be inaccurate, counterproductive, or unhealthy.
    a) A relatively few guidelines, perceptions, or goals can slow, corrupt, or prevent people's ability to competently address the life tasks.
    b) People's unique set of life style convictions have inherent biases that may lead to disagreement in subjective topics without a clear answer.
    c) When life style convictions are unsuitable or too rigid (e.g., prohibit goal achievement or healthy actions), people can experience symptoms.
11) Fictional finalism is a vague, guiding image of how life should be.
12) People may be frustrated or depressed when they perceive others or themselves as not on track or unable to live up to their fictional finalism.
13) The conceptions of a "Real" man and a "Real" woman are each person's subjective, unique definition of ideal characteristics that enable competency, but can include unhealthy thoughts, actions, and goals.
14) "Real" man and "Real" woman concepts may be used as something to strive for, as well as a yardstick of worth of self and others.
15) PROBLEM: The greater the difference between the self-image (how people see themselves) and the self-ideal (what they believe they should be) may be associated with increased inferiority feelings and a sense of being overwhelmed, ineffective, unintelligent, frustrated, or defective.
16) PROBLEM: Depression and frustration may arise when industrious, kind, thoughtful people have unrealistic expectations of others who

do not act in the same manner and fail to reciprocate compassionate or dedicated actions.
17) People must adapt unhealthy life style convictions and goals to survive and thrive in evolving and fluctuating environments.
18) Symptoms are, in part, a feedback mechanism that lets people know that their way of thinking and goals need to change for them to be healthier.
19) Life style convictions can be tenacious.
    a) Early-learned life style convictions can be especially sticky and stubborn.
    b) Life style convictions that allow goal achievement may strengthen, survive, and thrive, whereas unproductive ones may go extinct.
    c) Life style convictions can stealthily and powerfully reinforce themselves by defining what's valued, devalued, and ignored.
    d) The force and frequency by which people adhere to their life style convictions testifies to their dominance.
20) CHALLENGE: As life style convictions shape each person's perception of reality, people can be notably reluctant to alter them.
21) CHALLENGE: Life style convictions are so intertwined with perception that it can be difficult for people to detect their degree of influence.
22) PROBLEM: People's desires and the means to achieve them can be so strong they can become immune to argument and overrule some genetically-predisposed goals (e.g., sex, eating, socializing) and imperil their ability to adapt and achieve healthy objectives.
23) Neuroses can be equated to personality disorders.
24) PROBLEM: People's life style convictions (think lines of software code) comprise their personality (aka software) and can be associated with symptom development.
25) PROBLEM: Neurotic depression may arise when a person's life style convictions unhealthily deviate from common sense and/or are so inflexible that they hinder or prohibit adaptation and goal achievement.

a) Life style convictions may become so strong and inflexible that the related symptoms can arise quickly, frequently, and intensely.
b) Dogmatic tenacity to one's life style convictions is a primary and defining characteristic of neurotic depression.

26) CHALLENGE: Software-based (personality-related) depression can look like a hardware issue (psychiatric concern).

27) SOLUTION: Therapy is metaphorically-akin to a "software update" that reduces neurotic symptoms by an accommodation of healthier perceptions, goals, convictions, etc.

Private Logic

Imagine someone late one night sitting in a restaurant that has an extensive dinner menu. The individual cannot order everything that's offered, as no one could eat that much. Therefore, the person must select what appears to be the most appetizing, reasonable, and best able to satisfy hunger.

In pursuit of those goals, the patron can order any combination of main entrée, side dishes, appetizer, dessert, and beverage. In fact, whatever the person chooses is allowed within the common guidelines (the menu), the thoughts, actions, and goals are completely acceptable, and no one is alarmed or bothered.

But imagine the customer asks the chef to try something that the restaurant never offered: To crumble up cooked bacon and mix it in with pancake batter to make bacon pancakes.

As it's not on the menu, and it's well after breakfast time, there may be some resistance to it. Still, the patron assures the chef that the recipe has been successful at home, and it would benefit the restaurant if the chef were to try this new way of making pancakes with readily-available ingredients. When nearby others intrigued by the concept of bacon pancakes express their desire to taste the suggested dish, the chef tries the new recipe.

So, even though the customer has a way of thinking and acting that's a bit different from the norm, after some communication and social consensus, it's permitted. If enough people prefer it, it may become a regular item.

Now think about how the menu can be equated to common sense. There's been communication and consensus of what should be allowed within that community (restaurant), and people value those goals (menu items) and related way of thinking, thus act in agreement with that common sense.

When someone chooses specific available menu items, it's one metaphorical example of what's known as *private sense* or *private logic*.

Stated differently, it's what that person deems logical, realistic, and able to attain a goal (i.e., satisfy hunger).

Private logic can exist within the common sense.

Yet, when the customer exhibited private sense that's different from common sense (e.g., bacon pancakes, wanting breakfast food during dinner time), it was met with resistance. But, with communication and consensus that way of thinking and acting was deemed permissible and, later, desirable.

Private logic doesn't have to overlap common sense to be realistic or beneficial.

Some people think in a way unlike – but advantageously ahead of – common sense, and eventually widely adopted (e.g., Einstein, Newton, Darwin, Galileo).

Private logic that's initially outside of common sense can become part of it.

But issues can arise.

Imagine if a person stubbornly demands something that isn't offered and/or the community (e.g., restaurant manager, chef, other patrons) finds that choice to be offensive and unhealthy (e.g., warm pickle juice mixed with a raw egg).

The restaurant manager, wait staff, chef, and other patrons may provide social feedback, (e.g., become frustrated, angry, criticize) and reject the idea *and* the customer (e.g., ushering the person out of the restaurant). Also, if the person consumed that mixture elsewhere there might be bodily feedback (e.g., upset stomach, vomiting) from obeying unhealthy private sense.

It's problematic when people have private logic that's socially impermissible or exploitive, inflexible, and associated with symptom generation.

This analogy relates to neurotic depression.

People always think and act in accord with their private logic.[221] You may think of it as each person's:

- Unique set of rules, goals, perspectives.
- Evaluations of self, others, and the world.

- Belief of what's required to be competent in life.
- Set of convictions that may be illogical, unrealistic, and biased.
- Convictions of which the person may not be fully aware.
- Convictions which may not make sense to others.
- Ways of thinking and acting that serves to fulfill fictional finalism.

Fitting this into an existing metaphor…
- Personality is a general term, as is software.
- Life style convictions comprise personality, similar to how lines of software code constitute software, or an operating system.
- Private logic is more analogous to specific applications or subroutines.

While private logic can be in accord with common sense and public welfare, when private logic unhealthily varies from social interest, it can be related to neurosis.[222]

Thankfully, therapy can help people become aware of their problematic private logic and bring it more in agreement with healthy common sense.

Yet, the above factors make change exceptionally challenging.

First, *people can be unaware of their private logic and how it relates to symptom development*. They may misperceive other factors as causing their depression (e.g., genetics, hormones, other people).

For example, when people believe that they're always entitled to get their way, they may be confused and saddened when others no longer associate with them. As they cannot see their role in symptom creation and being alone, they're more likely to blame others…which can protect their self-esteem and status, among other things (e.g., "I don't know why my daughter doesn't talk to me after all I've done for her. She's so selfish and disrespectful!").

Being unable to detect how their private logic is related to depression, people can be less likely to seek psychotherapy, thus may perpetuate their symptoms.

Second, when private logic conflicts with common sense, people tend to prefer and abide by their private logic.[223]

In other words, there's a bit of "home-field advantage" or "dealer wins in a push" bias when it comes to a conflict between private logic and common sense. When dysthymic persons don't accept common sense and others' feedback, they're likely to maintain unhealthy private logic and related symptoms.

Third, as private logic constructs each person's perception of reality, they may be reluctant to loosen their grip on what they believe is truly real.

Fourth, given the importance of goal achievement and people's belief that their private sense will enable them to attain their objectives and grant a sense of competence, belonging, etc., they may cultivate unparalleled faith that their private logic is superior.

Given that point of view they may fortify and perpetuate their beliefs rather than surrender them – *even when they're inaccurate and counterproductive*. After all, they believe their sense of reality is correct.

So, like loss-incurring gamblers in a casino convinced they have a foolproof system, people can costly conclude that it's merely their impatience or infidelity to their private logic that's the issue.

People can rationalize and make excuses which enable them to maintain their allegiance to their belief system and vow to be more devoted to their troublesome private logic.

Next, imagine an individual with the following private sense:

"If I'm a good person, fair, honest, accurate in my understanding, and avoid unpredictable situations and possibly dangerous behavior, then I will be successful. I can get a career which will allow me to be self-sufficient, find someone with whom I can start a family and spend the rest of my life, have many supportive friends, save enough for retirement, and enjoy time with my grandchildren."

The above appears healthy, reasonable, in accord with reality as well as common sense, and a demonstration of social interest. Yet, think about how this private logic can be related to depression.

Well, consider how failing to live up to these goals or believing that others share the same objectives and ways of living when they don't (e.g., the spouse doesn't want children or spends excessively) can be associated with symptom development.

This is only the beginning.

People can have private logic that's linked to persistent depressive disorder in other ways (e.g., when their private sense maintains low self-esteem, prompts retreat from socializing, leads to an avoidance of sex and relationships).

Personality powerfully influences how people address the life tasks and can be related to symptom development when private logic is illogical, ineffective, inflexible, exploitive, and/or unhealthily diverges from common sense.

The Quest for Common Convictions

To understand dysthymic disorder, you must know the life style convictions symptomatic individuals share that's related to developing depression. However, given humanity's kaleidoscopic composition, this appears to be a task so massive it could crush an elephant.

For example, consider how dysthymic persons…

- Have assorted careers.
- Face different stressors.
- Enjoy a range of hobbies.
- Belong to various political parties.
- Attained different levels of education.
- Are married, divorced, widowed, and single.
- Hold diverse religious affiliations, or none at all.
- Are influenced by countless social factors and experiences.
- Do not have identical sets of convictions with other people.
- Live in urban and rural areas in various countries around the world.
- Maintain life style convictions that are intriguingly and influentially tethered to each other in incalculable ways.
- Have innumerable (and often gently evolving) convictions that make it impossible to completely record all of just one person's set of convictions, let alone those of every person with neurotic depression.

Yet, to crack the code that can demystify neurotic depression it's crucial that you know those common convictions. And you may consider the following to get started.

Why is it when you buy a book from Amazon.com or a song through the iTunes Store you are given a list of other publications or music purchased by those who bought the same book or song as you?

As people are goal oriented, it may be best to begin from that point.

While you recognize that your desire is to buy a book or song, the vendors' objective is to make a profit by selling items to consumers such as you.

But merchants have a problem to overcome.

They must figure out what you'd be interested in buying. If they knew your life style convictions regarding books and music, they'd have information that would allow more accurate recommendations.

Regrettably (at least for them), they do not.

When you think about it, the only data vendors have is *other* customers' purchase histories. But what good is that? These individuals aren't you. Also, as the diversity of online shoppers around the globe is vast and complicated, the challenge becomes all the more complex.

Retailers use other consumers' information to display items that you're likely to buy. Simply, they attempt to identify common convictions (comparable interests) to make more accurate recommendations to you. In other words, they use other customers' purchase histories to find patterns of preferences.

So, for instance, if you're browsing a medical thriller novel by a particular author on Amazon.com or a Chicago blues song by a specific artist through the iTunes Store, the purchase history data of all the other shoppers who bought that novel or song, as well as other books and music, can now be used as a reference in a process known as collaborative filtering.[224] Merchants shrewdly bet on these similarities to profitably increase sales.

Trends become evident with a large enough sample of individuals.

Knowing that, you can see how with a statistically significant number of symptomatic persons, the common life style convictions associated with dysthymia become increasingly apparent.

"But," you may say, "people's psychological symptoms are far more complicated and nuanced than buying a book or song."

True, but consider what the following suggests.

Deoxyribonucleic acid (DNA) is staggeringly intricate and incredibly influential. However, DNA's genetic encoding pattern consists of only four nucleotides (adenine, cytosine, guanine, and thymine).[225] Note as well that slight changes in the placement or appearance of any one of those compounds can create significant differences in presentation or ability.

Or, think about how your phone's software controls everything it does (e.g., how it presents, the way in which it perceives and interacts with you and the world). Yet, 100% of this massive, complicated computation and functioning is accomplished by only processing 1s and 0s. That's it.

Complex things have simple beginnings or components.

But there's one more thing you need to know.

Your phone's central processing unit – its "brain" – isn't aware of the quality, accuracy, efficiency, or purpose of the code it processes. The hardware uncritically and completely obeys the software's commands.

With that in mind, you realize that your phone can run a word processing program, web browser, or photo application as impartially and indifferently as it would…*a malicious virus*. This hints at how life style convictions relate to persistent depressive disorder.

Now, think about how…

- It only takes a tiny portion of the genetic code to define Parkinson's disease, hemophilia, Down syndrome, skin or breast cancer.
- Just a small percentage of an app's software code can make it a virus that destroys data or controls how a computer or phone functions.

Likewise, it can only take a fraction of people's entire array of life style convictions for neurosis to rise. Consequently, you don't need to know every one of a person's life style convictions, or all of the common convictions among dysthymic individuals, to comprehend symptoms. That would be an impossible task.

You only need to know the relevant ones.

The Depression Code

Knowing that trends become evident with a large enough sample of individuals, that complex things have simple components, and that it doesn't take many factors to significantly impact functioning, you only need to discover the convictions relevant to persistent depressive disorder to understand how it can arise, survive, and thrive.

So, even though dysthymic persons are remarkably diverse, the data of many individuals has revealed a distinct and finite collection of pertinent life style convictions. These are general rules for living (*modus vivendi*), rather than specific directives (*modus operandi*), to be competent in the life tasks.

There are three identifiable life style convictions related to neurotic depression.[226] To be concise, I'll call them The Depression Code.

The Need to Get

Initially, the need to get risks sounding as deplorably greedy and selfish as a corrupt politician, but that's an insufficient – and rather inaccurate – perception.

Consider how the need to get can be a healthy and beneficial rule to live by.

Just think about how to survive and thrive people must get things such as food, clothing, shelter, and safety. In addition, the need to get may guide people toward getting into better shape, getting a job, earning a promotion, accruing wealth to afford a family and have a robust retirement account, vacations, or desired objects (e.g., a summer cottage). Also, note that people can get intangibles, such as social skills, education, guidance, assistance, friendships, justice, and relationships.

Next, consider how a doctorate, honor society placement, and an Olympic gold medal demonstrate that the need to get can be socially valued and celebrated, as well as demonstrate competency or superiority.

In addition, consider how the need to get can be used to help others (e.g., being good at fundraising, drawing attention to unfairness or discrimination).

Given that life style convictions can be somewhat general, recognize that getting can vary in degree or topic. In fact, there are countless desires and ways in which to acquire whatever is sought.

The need to get can be a personally and socially valuable guideline that successfully enables competency in the life tasks as well as maintains or elevates self-esteem, status, acceptance, and a particular style of living.

But you also need to know the flipside of the need to get.

It's the aversion of losing anything of value or fear of being cheated.

This too, can be healthy and beneficial. Just reflect on how it may increase vigilance for those situations in which one can be exploited, say being taken advantage of financially (e.g., being alert for deals that sound too good to be true, or cautious of a shady sales person), or behaviorally (e.g., not put into a position of doing most or all of someone else's work).

The need to get – and the associated need to avoid being treated unfairly – can fuel people's efforts to make life fair and just for themselves and others (e.g., to be on the side of mistreated or neglected victims and those who don't have a voice, protecting those that cannot defend themselves).

To better understand neurotic depression, you must know that dysthymic individuals hold the need to get life style conviction.[227] For instance, you can see how people may become sad when they or others are exploited, or when they fail to achieve what they believe they need to get (e.g., a promotion).

The Need to be in Control

Control, in and of itself, can be a healthy thing. For instance, it's beneficial – for you and others – to be in control of the car when you're driving. Indeed, people want to be in control of countless things (e.g., their spending, eating).

Yet, there's a common misconception about control.

People often state that they need control, as if it were a goal. To see why that's incorrect, think about those who say they have the goal to exercise. It sounds right, but it overlooks their final objective. Actually, they use exercise *to fulfill another desire*, such as improve their health and live longer, prepare for a marathon, or lose weight.

Now consider the following to figure out why when people say they want control it's actually an intermediate objective.

- Being in control of a car avoids a nasty accident.
- Controlling spending avoids bankruptcy and losing one's home.
- Governing one's sex drive can evade divorce, shame, and disease.
- Restraint with one's eating can avoid obesity, a heart attack, high cholesterol, or having to buy new clothes.
- Influencing others' opinions and actions may maintain acceptance, gain status, keep the peace, or ensure employment.
- Being in control of one's temper (physical presentation, language) can avoid embarrassing situations or something more severe (e.g., arrest).

Control isn't a goal. It's a way to achieve one or more objectives.

People want to control an outcome. For example, they want to pass an important test or have a party go well. These are good and reasonable things.

Dysthymic persons often state they want control, as if it were an objective. Yet, you now know to examine that statement further. "In control of what?", "And, for what outcome?", are questions that may uncover telling clues.

To survive and thrive, people need to control various things, such as their muscles, speech, and various tools. Sometimes, people attempt to control others to achieve beneficial outcomes, as is the case with social cooperation say within a sports team or business, or while organizing a surprise party – all in an attempt to achieve another goal (e.g., win a game, provide goods or services, honor a friend whose birthday it is).

Perhaps the most common illustration is how parents attempt to control their children – say, to have them eat well, do their homework, develop a good work ethic, and acquire social skills – so they can grow up to be responsible, well-adjusted adults who are better able to competently fulfill the life tasks.

The need to be in control can prompt positive characteristics, such as planning, organization, reliability, problem solving, kindness, and leadership.

But there's something else related to the need to be in control.

Consider how you might feel being on the edge of a steep cliff without a safety rope, riding on a serpentine roller coaster unrestrained, or in the back seat of a car recklessly careening down an ice-slicked road.

It's likely you would experience the fear and bowel-agitating panic associated with being out of control. This feeling can be exponentially multiplied for those who need to be in control…and related to the following.

Those with the need to be in control try to avoid situations they fear may go out of control (physically, socially, financially, emotionally, etc.).

This can be a healthy pursuit, as they may evade or escape situations that are risky and imperil their ability to control (e.g., walking down a dark alley and getting robbed or raped, going to a casino and gambling away their money, using mind-altering drugs or consuming too much alcohol, breaking the law and being imprisoned, being in a sexually-tempting situation).

With this in mind, think about what characteristic is intimately intertwined with the need to be in control.

Hint #1: What do people seek so that they can stay in control?

Hint #2: Why may those who need to be in control think "all the time"?

Hint #3: Why might they extensively research things, events, or people?

Those who need to be in control, *want to remain in control*. Accordingly, like a substitute teacher facing a notoriously-rowdy class, they can hate being taken off guard or otherwise have things go in a way they didn't expect.

Those who need to be in control strongly desire predictability.

When people can anticipate or calculate what may occur, they can plan and can take measures to preemptively avoid a loss of control. So, not only are they unsurprised if a situation arises which would otherwise risk a loss of control, they're prepared to address it, thus remain in control.

There are countless ways people seek predictability. Here's a few.

- They may frequently think about how to prepare for something that's coming up (e.g., a family vacation and all of the related details such as hotel, transportation, necessary items) so that they can control for an unanticipated demand or emergency.
- Always carry extra money to control for an unexpected expense.
- Do exhaustive research (say on medication, a TV, car) to anticipate and avoid any loss of control (e.g., thoughts, emotions, device failure and related expense, being stranded on the side of the road).
- To control their health, they investigate medical conditions, their prevention, and take action in accord with their research.
- They may not drink to excess (or at all) or use illegal or legal drugs (e.g., antidepressant medication) because they don't want to hand over control to a substance or they fear the unpredictable loss of control (e.g., of muscles, of what they say, becoming "hooked on drugs") if they were to get drunk or high, for instance.
- Frequently ask questions to assess future situations, such as "Who is going to be at the party?", or "What's the weather going to be like?", and then attempt to control for an undesirable outcome.
- Have familiar, consistent routines that do not pose the risk of an unpredictable problem (e.g., morning rituals, only park in one spot at their job, always workout in a specific order of exercise, eat the identical lunch every day and in the same order).
- Read their horoscope daily.

- Be convinced they have premonitions or some other supernatural power of predictive clairvoyance and use their feelings to guide behavior.

Dysthymic individuals believe that in order to have a place they must be in control.[228] This life style conviction is strongly related to the need to have predictability.

The Need to be Good, Perfect, Right

To start, consider how being good is a praiseworthy characteristic that can be used to healthily and productively achieve diverse objectives.

For instance, dysthymic individuals can believe that being a good…
- driver will allow a safe driving record and low insurance rates (or steer clear of speeding tickets and accidents).
- citizen can enable them to earn credibility, a respectable reputation, and garner community accolades (or avoid criminal investigation).
- spouse guarantees a loving and respectful lifelong marriage (or can sidestep divorce or an affair).
- employee will ensure long-term, stable employment and frequent promotions (or prevent unemployment and poverty).
- parent will lead to having happy, healthy children (or not raising spoiled kids who are endlessly troublesome and disappointing).
- friend will assure lifelong support, connection, interaction (or evade painful loneliness).
- person may be the key to what they consider the greatest reward for goodness: An eternally-rewarding afterlife (or avoid a hellish one).

Being good can help people successfully address the life tasks.

While living according to what's acceptable and encouraged within common sense is applauded, some people can go well above what's acceptable. Those with persistent depressive disorder can be remarkable friends, terrific siblings as well as dutiful and respectful children. In addition, they can be hilariously funny and virtuously altruistic. Some can be extraordinary people pleasers who generously and compassionately act with others in mind.

As you've probably guessed, the components of this conviction are related. *Dysthymic individuals can want to be so good as to be perfect.*[229]

Some people with neurotic depression can be strict perfectionists, in one or more areas. This can be their way to ensure attaining desirable outcomes.

The related need to be right, should make sense as well. It wouldn't be good to others, for instance, to give out inaccurate advice or information. The need to be right can manifest itself in various and beneficial ways. For example, dysthymic individuals can know the exact way to do something, maintain correct facts, figures, and details (and either seek it when they don't have it or refrain from giving it if they don't know the right answer).

While people can pride themselves on being right, they might readily receive others' correction (which seemingly can risk acceptance and status) because it allows them to be right in the future.

Revealingly, people often seek out those who have the need to be good, perfect, right. For example, if you were having heart surgery, you may want a well-intentioned perfectionist who knows exactly the right way of operating on you. Likewise, there's a benefit from employing a thorough and meticulous accountant, who knows the correct way to fill out the necessary forms.

Those with neurotic depression have the need to be good, perfect, right.[230] While they can use this conviction to successfully fulfill the life tasks, when can it be related to symptom development?

Well, just think about how people can get upset with themselves if they acted in a manner that they deemed morally bad, imperfect, or factually incorrect.

Dysthymia is commonly attributed to a hardware concern – and, indeed, there are legitimate psychiatric causes. However, to comprehensively decode persistent depression, there needs to be an acknowledgement and awareness of how people's personality is associated with symptom creation and perpetuation.

The Depression Code convictions – the need to be in control, the need to get, and the need to be good, perfect, right – are related to developing and maintaining neurotic depression.

Those who live by the Depression Code to address the life tasks can be amazing individuals who regularly achieve various personally and socially-beneficial goals. Unfortunately, those convictions can lead to developing symptoms.

Problems with the Need to Get

While people need to get exercise, pursuing it with the unrestrained fight of a gladiator can lead to symptoms (e.g., exhaustion, pulled muscles, knee injury). The same goes for eating. People need to get food, but overeating is harmful.

When people get too much, or excessively strive to get, their actions can be stressful, counterproductive, and related to symptom development.

But neurotic depression is far more complicated than just overfilling the need to get and overconsumption of otherwise healthy things.

Dysthymic individuals can try to get *unhealthy* things.

Just think about how people can get…

- Revenge.
- Undue attention.
- Others to do their work for them.
- Material items far beyond what's necessary.
- Money to the point of harming others' well-being.
- Unrestrained levels of pleasure (e.g., drug use, random and perilous sex).
- The most number of friends (online or in real life), but often of low or questionable quality.
- Special treatment (e.g., sympathy, exemption from responsibilities, disproportionate choice of how family time and money are spent).

The need to get conviction can be unhealthily boundless, as there are countless harmful things people can try to get. Moreover, it can include the pursuit of self-centered, exploitive, and harmful goals as well as personally and socially-detrimental ways to get them.

Also, the need to get can require the most sensitive perception to detect. For instance, people may believe they *must* take advantage of a sale, even though they don't need the item. Those with the need to get can seek tangible things (or that which allows their grasp), or intangibles, such as power, status, or attention. They may expect others to donate time, money, goods, and services.

Some may have a significantly unhealthy sense of entitlement.[231]

Consider how believing they're entitled enables them to get without concern for others' well-being, or restraint – say unfairly giving themselves permission to get. For example, people can believe they're entitled to respect as they're older or hold a position (e.g., in-law), but treat others disrespectfully (e.g., a one-way hurling of insults from behind a protective shield of respect).

The need to get conviction may lead to taking on the "I am what I have", "I want what I want and I want it now!", or "The person who dies with the most toys wins" mentality. When people believe they must get, but do not (for whatever reason), they may perceive themselves as hopeless failures, sadly flawed, or cosmically unfortunate.

Those who have the need to get conviction may equate their self-worth with their ability to get, as well as the quantity and quality of their possessions.

The need to get can be a restless and unending pursuit, therefore, unsettling, overwhelming, unsatisfying, and self-defeating. After all, it's something people use to achieve countless goals. When they cannot succeed living by their chosen conviction, they may feel doomed to failure and rejection.

Yet, some might protectively, but problematically, avoid responsibility for the outcome or gloss over the need to improve their conviction.

Dysthymic individuals believe that in order to have a place they must get, and if they don't, they can get angry and/or deem life and others as unfair.[232]

Also, keep in mind that with the need to get, there can be the fear of losing anything of value or being cheated in some way.

While this can be employed in a healthy manner (e.g., avoid unemployment, divorce, bankruptcy), it can also be related to symptom development when individuals rigidly obey the life style conviction or employ it in a harmful way (e.g., get upset or sad when a new person joins the existing group of friends out of jealous concern of losing those cultivated connections).

Those with the need to get can perceive loss as movement away from competence and toward inferiority.

When people get ruthlessly swindled, for instance, they may perceive it as their inadequacy in addressing the life tasks and see themselves as less competent (and perhaps fear others hold the same perception) – which imperils self-esteem, status, acceptance, and their way of living.

Those who need to get may take loss personally.

This can happen needlessly and disproportionately. For example, whether it's by chance (e.g., storm, traffic lights), or their own mistakes. A loss can be of any category or size for those who need to get to experience symptoms – their perception, rules, and goals influence whether depression develops.

You can see how those who need to get may experience envy, frustration, and depression when others are more successful at getting.

For example, when...
- A sibling gets more parental attention or emotional support.
- A co-worker gets selected for a promotion.
- A fellow diner whose imperfect meal was taken off the bill.
- A friend made more money at a garage sale or in the stock market.

Those with the need to get can be unnecessarily and painfully competitive.

Also, they may continually assess and evaluate as a way to ensure getting. For instance, they may keep in touch with friends who have access to places or people, or who can enable some benefit down the road.

Or, if they believe their friends will become a liability (e.g., need emotional support, request help to move, ask for a loan), they may abruptly end those relationships – or ghost those connections, having them dissolve and dissipate from neglect. Unfortunately, the need to get can lead to mistrust and sabotaging otherwise nourishing social connections.

Those with the need to get may view others for the benefit they provide, with vigilance and suspicions rising upon the thought of the costs they may inflict.

To prevent forfeiting something of value, some hoard money, objects, or even insignificant things, with the belief that they may be worth something one day or needed at a critical moment. People may hold onto various items of little to no value – often at family members' displeasure – such as tattered childhood playthings, knickknacks, etc., with the unassailable rationalization that if they did get rid of them, that's *exactly* when the value will skyrocket, of course.

Distressingly, those with the need to get conviction may never satisfy their hunger or be at peace. They can believe that their getting, or what they have, is unendingly imperiled, creating a perpetually-unpleasant existence.

Recall that the need to get can guide individuals to healthy practices, such as being sensitive to unfairness as well as having empathy and energy to protect and advance those deemed as powerless or unfortunate victims.

Consider how that keen ability to detect injustice can relate to depression.

People can spend a significant amount of time contemplating what's fair, and who should get what. However, life is often brutally unfair. Therefore, those with the need to get conviction can maintain an *unrealistic expectation* of fairness, which can end in frustration, anger, or sadness.

Those with persistent depressive disorder can (unrealistically) expect everyone else to act just like them (e.g., in a good, fair, supportive way). Indeed, they commonly project onto others what they should do, or should have done, for instance, "Well, if it were me, *I* would have ____."

Note as well, those sensitive to the slightest infractions of fairness may overdo it. They can misperceive themselves as supernaturally unlucky or helpless victims, who are abused, mistreated, exploited, etc.

This may intensify a depressing mindset. Also, such notions may prod their suspicions of others like a hornets' nest; stirring a swarm of stinging thoughts and rapid, protective distancing that can ultimately magnify symptoms.

Problems with the Need to be in Control

While the need to be in control can be positively employed in countless ways, consider when it can become problematic.

First, symptoms can arise when the need to be in control is greater than what's possible or healthy.

The need for control can extend to all matters. That is, some people want superhuman power and authority. Yet, others want to control everything within a narrow section of life (e.g., home, driving, money, vacations, appearance, TV remote, sex). While the means to gain and use control are countless, the need to be in control becomes problematic when people want too much – *even when it's well-intentioned*. For instance, concerned and kind parents who want to control their adult children's spending, dating, diet, career, or friendships.

Second, *the need for control can be related to depression when people are powerless*. When certain matters are uncontrollable, people can feel depressed and hopeless (e.g., grave medical illness, war, other people's wrong impression, widespread famine, friends with poor self-care, relatives who lead a dangerous lifestyle [e.g., risky sex, drug use, obesity]). This is reasonable and relatable, but the experience can be *exponentially* more stressful for those who need to be in control and are intensely reluctant to be at the mercy of anyone or anything.

Dysthymic persons live by the need to be in control conviction and may strive to be in charge of everything, as well as know as much as possible.[233]

Yet, in life there are very few things under any one person's command. Therefore, those with dysthymia will most assuredly encounter situations outside of their jurisdiction. When they're unable to, prohibited from, or fail to fulfill their strict self-imposed need to be in control, they may feel depressed, angry, and anxious, as well as suffer psychologically-based physical symptoms.

Third, those who have an intense and inflexible need to be in control can develop a disproportionate dread of matters becoming unmanageable. This can explain why dysthymic individuals can become symptomatic when everything is fine. They're expecting situations will go out of control – whether or not they actually do – and perhaps fear they'll be powerless to manage them.

Knowing that the need to be in control is related to neurotic depression increases people's awareness of what to assess and can helpfully guide treatment. With that in mind, consider what areas of functioning when threatened with a loss of control can be associated with symptom development.

Those who need to be in control have at least one of the following four major fears.

1. *The loss of physical control.*[234] For example, paralysis, being physically restrained, erectile dysfunction, being in an aircraft they can't deplane, being on a long bridge or thruway they can't exit, in a meeting they can't leave, wearing tight-fitting or restrictive clothing, or immobilized in an MRI machine, or elevator, on a high cliff, balcony, or ski lift.
2. *The loss of psychological control.*[235] For instance, loss of sanity (dementia), losing control of their emotions (e.g., getting angry, crying).
3. *The loss of control over how others perceive them.* For example, undue worry about how people may judge them negatively.
4. The ultimate loss of control, death and dying.[236]

While none are a day at the beach, those with persistent depressive disorder can be exceptionally prone to develop symptoms in regard to these feared conditions. For example, they may become saddened, anxious, worried, or demotivated upon contemplating or enduring situations in which they lose physical control, psychological control, influence over others' judgments of them, inescapable death – or the process of dying, in which they inescapably and helplessly witness their own demise.

Often, dysthymic individuals can significantly fear losing control of their physical movement that would create a more dreaded condition.

For instance, they may avoid being on a high cliff or balcony out of worry they'll lethally lose control of their movement (e.g., lose their grip, misstep [perhaps due to a feared moment of inattention]). Some consider jumping (e.g., from a ski lift), not in a suicidal way, but to escape the intolerable, tension-saturated sensation of powerlessness and imprisonment.

Those who like control may fear and rebel against anything in which they feel controlled.

This may occur when they're unable to leave a situation, for example, an important meeting, on the thruway or lengthy bridge. (Supportively, there are services that drive people across long bridges when they're fearful of being restricted and powerless to escape them.[237]) Some avoid tightfitting clothing as they can feel trapped, as if in some modern-day, off-the-rack straightjacket.

Dysthymic persons can take specific action to deal with dreaded situations.

For example,

- *To avoid the loss of physical control*, they can be overly concerned with safety (e.g., avoid plane travel by deeming it unsafe, take the stairs instead of the "questionable" elevator, go the long way around rather than feel trapped on the thruway, a long bridge, or in a tunnel [which they may view as threateningly unstable or prone to a cave-in]).
- *To prevent the loss of psychological control*, they can go to great lengths to not cry or get angry. Some ensure no one sees them do so, say by crying in the shower or car, or getting angry only when alone. Some are unstoppably drawn to medication that sedates their emotions. Others may avoid situations that risk the loss of emotional control, such as watching horror movies, or by making sure that no one ever throws them a surprise party. Despite painful and disruptive symptoms, recall that some people may avoid antidepressant medication – or any legal or illegal drug – out of fear it will lead to a loss of control, then be at risk of saying or displaying something they want

protectively camouflaged and covered. Some relentlessly and inextinguishably take medication or engage in activity they believe will prevent cognitive decline.

- *To avoid the loss of control over others' perception and judgment*, they may ruminate about how they did something that made them look foolish – perhaps weeks, years, or decades earlier – and beat themselves up verbally and emotionally as they swear to never do it again.

  They may restrict their existence to specific unthreatening domains to avoid others, such as remain at home rather than go to a party where they cannot control what others think, say, or do. Often, dysthymic persons only socialize with family members, as they're predictable and far less likely to abandon them, as they fear others will.

- Death and dying are inescapable. Yet, some forcefully strive to avoid an early death (e.g., dutifully exercise and eat well), hold anesthetizing views of death, ignore their weight and unhealthy habits, or limit the risk of powerlessly seeing their expiration (e.g., euthanasia, a Do not resuscitate order). Some, as the saying goes, "Tiptoe through life just to arrive safely at death", never fully living out of fear of hastening their end.

Given dysthymic persons' intense need to be in control, when they feel out of control, or fear losing it, they can experience symptoms such as, frustration, anger, feeling overwhelmed, anxiety, and depression.

Accordingly, they may resolutely avoid arenas and persons they're unable to control. People can become depressed when things or others are outside their influence. For instance, when friends and family don't heed their (unasked for) advice, when they're unable to predict or control their depression, or powerless over situations that risk failure, ridicule, rejection, or abandonment.

Those with dysthymia may most fear a loss of control when they face a challenge they don't believe they're capable of successfully completing.[238]

This should be unsurprising as it underscores the importance of human goal orientation and competency. Note that avoiding failure can protect their self-image, how others perceive them, as well as not imperil their way of living (e.g., ask for a raise and get fired).

Those with the need to be in control usually avoid unpredictable situations. (Though there are exceptions, e.g., games, sports, movies, or when they're in charge of creating it.) With that, consider what's often related to this desire?

Hint #1: Predictability.

Hint #2: Anxiety often co-exists with depression.

Dysthymic persons may be hawk-eyed and alert for anything that may go out of control, to the point of generating and maintaining symptoms.

Sometimes, it's rather innocent looking; for instance, those who "cannot stop" watching the news, stock market ticker, or weather report.

Now consider why those things are particularly telling.

By perpetually monitoring for even the slightest changes, they can have greater predictability and a head start by sensing when things are about to go out of control and swiftly jump in and take charge of a situation (e.g., take their money out of the stock market, stock up on food and firewood). While awareness and quick action can be healthy, dysthymic individuals can persistently suffer given the duration and intensity of their vigilance.

Some constantly check others' online status, pictures, and recent events, to guard against something going awry. For example, they may suspiciously, yet stealthily keep tabs on their adult child's social life, how long their friends have been offline, or that many co-workers have recently taken jobs at other companies (and perhaps fear upcoming layoffs, a takeover, or bankruptcy). Perpetual concern that things will go out of control can be dreadfully depressing and anxiety-provoking.

Those who need to be in control may go to extraordinary and unhealthy lengths to have predictability in their environment and the people in it.

The need to be in control, desire for predictability, and related wish to avoid chaos can lead people to regularly search for stability and predictability.

With that in mind, consider why dysthymic persons might do the following.

- Stay at a dull and repetitive job.
- Dutifully celebrate anniversaries.
- Maintain a difficult and distant marriage.
- Have an unbendable, yet inefficient, routine.
- Go to an unsurprising store with an anemic selection.
- Be committed to making sure a family tradition is passed on.
- Only go to one restaurant and/or order the same meal every time.

People can attempt to control otherwise unpredictable and undesirable outcomes they fear would occur if they made a change.

Think about how this might lead to unhealthy actions and symptoms.

By keeping a dull and repetitive job, they're familiar with the obligations, setting, and people. This can prevent unnervingly-unforeseeable changes such as, applying for a new job or being in an unfamiliar environment. However, this way of living can be claustrophobically limiting and depressing. Change can lead to unpredictable circumstances and the fear that things will be worse (e.g., "Out of the frying pan and into the fire" type of thinking).

Celebrating anniversaries can be a good and healthy practice. Yet, also recognize that anniversaries highlight and reinforce the importance of stability. Those with persistent depressive disorder may feel hurt, annoyed, or saddened when others do not observe anniversaries with equal reverence and dedication.

By sustaining a difficult marriage, people may sidestep being thrust into an unpredictable singles' world, at the mercy of others' whims, feelings, sexual history and health. Also, they get to control where they live, their finances, and perhaps what other people think of them. But, they may come to a sad realization that they're merely keeping a dying union on life support.

Maintaining an inflexible and ineffective routine ensures predictability. But, people can feel drearily shackled to an unsatisfying way of living. Also, they may experience agitating anxiety whenever their revered and rigid routine is interrupted, postponed, or imperiled.

Going to a small, but familiar store allows people to have predictability and control. For instance, avoid being caught off guard or having to unexpectedly spend more time than desired (e.g., "I know exactly where everything is", "They all know me by name"). But, there's likely a hit to their quality of life when they restrictively have to go without items the store doesn't carry.

Keeping family traditions can be healthy, kind, reassuring, and enjoyable. Doing so can be particularly desirable for those who need to be in control and want predictability, as it enables them to keep with what's familiar.

However, the desire to maintain traditions can be problematic for some.

Traditions can become outdated, irrelevant, and have uncertain origins, which breed doubt of why they're maintained. Consequently, family members may not care for them. This can lead to friction, alienation, or abandonment.

Also, consider how people may control their children, siblings, and spouse by mandating that they obey the custom (e.g., "It's been a tradition forever. We *have* to stick with it!"). Traditions, like anniversaries, fortify the importance of stability. Paradoxically, some dysthymic persons may feel frustrated and sad when – in an attempt to be good, for instance – they've imposed self-restrictive behavior (e.g., "I don't want to make a fruitcake for everyone and send them out, but it's been a family tradition for a century").

People who only go to one restaurant and/or order the same meal every time (perhaps saying, "It's my favorite" and "I know it's good") is revealing. While they may genuinely like the meal, they prohibit themselves from enjoying the remainder of the menu. Some fear the unpredictability of getting a bad meal.

Those with the need to be in control and strongly desire predictability may endlessly attempt to be omniscient and omnipotent – wanting to predict the future and control outcomes. Sadly, even when these pursuits are well-intentioned, they invite symptoms.

In their planning and focus on future events, dysthymic persons can distract themselves from the present and erode their ability to enjoy the moment.

For example, they're not able to appreciate a meal because when they're ordering the appetizer they're thinking about the drive home from the restaurant.

Also, think about how life, others – and even themselves – aren't as predictable as desired. So, they may feel inferior when they fail to attain anticipated success or feel unloved when others do not act a certain way, for instance, friends or family who didn't show up to their graduation, aren't interested in their vacation photos, don't call as frequently as preferred, or only call when they need something.

The need to be in control is a problematic conviction that can endanger mood, self-worth, participation, acceptance, among countless other things, and related to psychological and physical symptom generation.

Problems with the Need to be Good, Perfect, Right

Think about how being good, doing something perfectly, and being right, is widely encouraged and applauded. For example, parents regularly attempt to instill this trio of linked and well-intentioned convictions in their children, businesses want their employees to have these characteristics, and governments and religions regularly define and advocate such aspirations.

Consider how the need to be good, perfect, right be related to dysthymia.

To start, recollect that common sense isn't very common.

With that in mind, contemplate the following:

- What's judged as good within a gang is much different from that within a family, workplace, or a political system (e.g., conservative republican, liberal democrat, progressive, libertarian).
- Some attempt to please all the people all the time, while others may ignore or annoy everyone while attempting to please their god(s).
- Individuals who strive to delight their tyrannical spouse while displeasing everyone else.
- There are those who attend to and support their co-workers or friends but disregard their families.
- What's defined as good, perfect, right changes over time and across geography.

That which is considered good or perfect or right is relative and subjective. It can be in reference to what society values, what a subculture respects, or what an individual cherishes. Simply, there are many ways in which people can comply with the need to be good, perfect, right. So, for instance, not knowing the "right" way of being good, or to whom to be good, or unaware of how to do something correctly or perfectly, can be frustrating and depressing.

As there usually isn't a clear, exact, and stable answer for what is good, perfect or right, such variable and uncertain alternatives can account for symptom development.

Next, what paradox can arise for those who have to make a decision; say, where to vacation, what restaurant to go to, or which movie to see?

They want to be a good person, pick the perfect restaurant or right movie to watch. Yet, how can they if they don't know the outcome?

They may try to reason it out – or seek unattainable clairvoyance. Alas, as they cannot predict a future outcome, they remain uncertain of the best decision. When dysthymic individuals don't know what to choose, they can paralyze themselves in fretful indecision – *especially if others are reliant on the decision.*

For example, they don't pick a restaurant or movie because they don't want to be held responsible for making a bad choice and disappointing or angering those who followed their decision.

There's a related problem for those who have the need to be good, perfect, right: *Difficulty lies in determining which decision is the better one between two (or more) good choices.*[239] It's fairly common for those with neurotic depression to wrestle with even the most trivial decisions and perceive them as monumental. In fact, they may view themselves as powerless, torn, and as victims.[240] This also can create cognitive deadlock.

But there's something else to keep in mind.

Those who need to be good, perfect, right, can be kind, thoughtful, generous, hardworking individuals, who are loyal friends, and sweetly dedicated in their relationships. They can be shining examples of the best humanity has to offer.

While those socially-praised characteristics would seem to protectively immunize people from depression, consider when symptoms can arise.

To start, recollect that an individual's symptoms often have a social factor.

Now think about those who need to be good may quickly and generously trust others. However, like a bank without a security system, being unsuspecting can leave dysthymic persons unable to detect and protect against those who prey on such readily available trust and kindness.

Next, consider what may occur when others *do not acknowledge or reciprocate* dysthymic persons' caring and industrious efforts.

As life style convictions can be strong and inflexible, people can be supremely reluctant to alter or surrender them – even when they're problematic.

So, rather than adaptively revise their convictions and actions, some magnify their faith in a losing strategy, and obey their belief with greater intensity.

Think about how can they do this without violating their mindset.

Dysthymic individuals may counterproductively justify their behavior and suffer in silence for "the greater good" (e.g., for the marriage, children, employer), and push themselves to their limits.

Many people in miserable marriages or insufferable jobs may rationalize perpetuating their circumstances because they don't want to see themselves – or have others judge them – as bad, imperfect, or as having made the wrong decision. While it's common for those who've spent time, effort, money, etc., to want a return on their investment, sticking with the same strategy can be like repeatedly buying a stock that continually loses value with the hope that it'll return to, and rise above, its former highs.

Now think about what can eventually happen.

Their fruitless self-sacrifice may lead to fatigue and failure.

Some escape their circumstances but employ the same unproductive conviction in another environment (e.g., new job, friendship, marriage). But, this often ends with a similarly-unsatisfying result.

Others depressingly retreat from situations with the firm belief that they cannot win – or have so exhausted their passion that they no longer wish to play.

People select how, when, and where they want to be good, perfect, right. Some strive to be the best employee, spouse, parent, friend, or as saintly as possible all the time. Unsurprisingly, those who want to be perfect in everything will inevitably fail, and depressing immobilization often follows.

Note the binary aspect of this life style conviction. It's seen in statements such as, "It's either perfect or it's wrong!" Accordingly, if people do something well, but slightly imperfectly, they can deem themselves as inferior and as failures…and fear others have come to the same conclusion.

When people have the need to be good, perfect, right, it can be an eternal and weighty cross to bear, as they believe that every day is judgment day, and they mere mortals, inescapably prone to transgression, and fallible. Eventually, the conviction is violated at which point they may catastrophize, for it's one thing to fail on others' measures of success, but to fail on one's own is particularly distressing. They cannot live up to their self-ideal.

Those with unrealistically high standards can become depressed.[241]

After a less than perfect thought, emotion, or action, they may…

- Dwell on how they broke their standards and believe they're failures, forever corrupted as they can never return 100% to perfection and their imperfection cast in stone (e.g., a student who didn't make Dean's list one semester wails, "It's on my permanent record!").
- Feel as though they cannot be good enough or consistently good. (With perfection as the compared benchmark, how could they feel competent?)
- Mercilessly replay the error in their minds.
- Engage in self-flagellation, making sure that they – and sometimes anyone within earshot – never forget their "sins" (e.g., "I'm beyond help", or "A person like me doesn't deserve to be loved").
- Believe they're doomed to an irreversible path towards an undignified or damnable condition, e.g., "There's no hope for me, I'm cursed", "I won't get into Heaven."
- Be perpetually penitent, apologize regularly and excessively.
- See themselves as the worst people in the world.

- Attempt to correct an unchangeable past, instead of adjusting to it.
- Berate themselves for some tiny transgression that occurred, days, weeks, or decades earlier.
- Consider themselves "as bad as everyone else" and no longer above criticism, which their perfectionism previously protected.
- Counterproductively retreat from people or opportunities.

The need to be good, perfect, right conviction when flexibly followed can empower a healthy pursuit of countless objectives, and enable people to achieve competency, acceptance, and status as well as maintain a certain life style.

However, this conviction also can be linked to developing symptoms such as ceaseless rumination, extreme self-torment, and an avoidance of situations in which they dread doing or saying the wrong thing, which can lead to a progressively limiting and unfulfilling life.

Dysthymic individuals may fear being bad, imperfect, or incorrect, with each believed to jeopardize goal achievement as well as self-esteem, status, acceptance, and their way of living. Moreover, when they either fail or are restricted from living up to their need to be good, perfect, right, they may feel inferior, incompetent, despicable, immoral, bad, broken, frustrated, saddened, unfairly victimized, or a range of other depressive symptoms.

Although the Depression Code convictions are not neurotic in and of themselves, when people obey them rigidly, they become increasingly prone to developing depression (particularly when they approach an unavoidable situation for which they're unprepared, such as a new job, marriage, socializing, or parenthood).[242]

You're now familiar with how a mindset that contains the Depression Code can relate to symptom development. But there's more than excessive and inflexible compliance to the Depression Code that's responsible for symptom development.

Inadequate and Troublesome Convictions

Full comprehension of neurotic depression requires that you grasp the massive influence personality has on people's goals, thoughts, actions, emotions, and symptom development. While inflexible obedience and/or excessive application of life style convictions can be guilty of stealing people's quality of life, these are only two of the usual suspects. There are more.

Insufficient and Incorrect

Imagine that you believe you know how to swim…but in actuality are a poor swimmer because you're unfamiliar with the proper technique. Accordingly, when you dive into a swimming pool, you're only able to tread water uneasily. Or, say you've precisely followed a recipe a relative gave you for baking your favorite cake. However, you wrote down the instructions incorrectly. In both situations, you thought you knew the right way of doing something, but you were sadly unable to achieve your desired outcome.

These illustrations are revealing metaphors.

When people have unsuitable life style convictions, misperceptions, faulty rules, insufficient knowledge, or incorrect beliefs, they may be unsuccessful in attaining their goals and develop symptoms.

It's easy to see how mistaken, naive, or incorrect knowledge or guidelines, for example, on how to be a good employee, economize, socialize, or be romantic and make love can lead to unemployment, bankruptcy, loneliness, or devastating feelings associated with failure and inadequacy.

Compounding

Take into consideration how your computer usually runs without a hiccup but may get bogged down when many apps operate concurrently, increasingly complex computations are executed, or numerous apps attempt to access the same data or device at once. This analogy provides a clue about human beings.

To start, note how people's maturational process brings an increasing array of life challenges. While infants have no responsibilities, as they age they must find solutions for the escalating obligations.[243] Various desires, responsibilities, and objectives – and their related convictions – rise rapidly; for instance, language acquisition, elementary school, high school, socializing, dating, sex, college, work, marriage, family, financial requirements, mounting physical and cognitive concerns, the loss of friends and loved ones, etc.

Life begins with simple and separated rules. But, it becomes overcrowded with a growing number of convictions, often of greater complexity, which magnifies the probability they'll problematically overlap.

Think of this like loving parents who get overwhelmed when all of their kids simultaneously demand attention and different action (e.g. diaper change, cleaning up a spill, tying shoes, feeding, reassurance, or coaxing from a hazard).

Knowing that excessive, competing demands can slowdown a computer or overload parents, you can foresee the following.

People can be overwhelmed by attempting to obey too many convictions or fulfilling taxing objectives – either simultaneously or serially.

With that in mind, consider those dysthymic persons who strive to do many things perfectly or be good to everyone (e.g., ideal employee, friend, sibling, and child to aging parents). The impossibility can lead to depressing failure.

In their well-intentioned endeavors, they may be crushed by an avalanche of personal rules and goals. For instance, to be a *good* friend they…

- *must* be the one to host the *perfect* party, *and*
- *control* the environment – as well as their friends' opinions of them – by
- making certain their house is *immaculate*, *and*

- providing more than enough *home-cooked* food, *and*
- serving a *wide variety* of edibles so that no one is left out, *and*
- offering *plenty* of sources of entertainment for the kids so they don't get bored or rowdy, etc.

As it's likely that they cannot do it all – or do everything well – they may feel as though they've failed, disappointed others, acted disrespectfully, or in some other way imperiled their self-esteem, status, acceptance, and way of living.

Conflicting

What happens when a computer attempts to obey incompatible lines of code?

The disagreement among instructions can lead to a computational quandary that slows or stops functioning.

Perceiving personality as metaphorically similar to software, it's easier for you to see how conflicting life style convictions can create logical dilemmas and lead to psychological and physical symptoms.

Simply, those with neurotic depression may not know how to resolve such disagreements or their convictions constitutionally forbid them from doing so. This can frustratingly imperil or prohibit goal achievement.

The incompatibilities among the Depression Code convictions are referred to as the Candy Bar, Cookie Jar, and Stoplight conflicts.[244]

Candy Bar Conflict

Imagine a dysthymic individual is about to eat a candy bar when the person's spouse approaches. To be good, the person kindly chooses to share the candy bar. Yet, upon dividing it, there's one piece that's bigger than the other.

Consider how this can cause a dilemma within the Depression Code.

If the person obeys the need to be good, perfect, right, then the bigger piece goes to the spouse. However, if the individual abides by the need to get, then then realize what's prompted.

Keeping the larger portion…and giving the smaller one to the spouse. (After all, just a moment ago the person had the entire candy bar.)

The two convictions are incompatible at that moment.

The person may pause and attempt to arrive at a pleasing conclusion. Yet, no matter the solution, *there must be a violation of at least one life style conviction*, which can lead to symptoms.

You may say that the dysthymic individual can pick to obey one conviction and then rationalize the decision (e.g., "It's my spouse who I love – whose favorite candy bar this is – therefore, I must give the bigger piece away").

Certainly, this can happen. Yet, life style convictions can be strong and inflexible. Consequently, people may create a double bind that strains incompatible convictions like a wishbone until one of them shatters.

In this example, the dysthymic person may feel sad that there was any debate about who should get the bigger piece, or there could be some unsavory and unbecoming annoyance or anger that the spouse appeared at that exact moment which created the dilemma, or not knowing which conviction to prioritize can immobilize the individual.

Cookie Jar Conflict

Picture a person who lives by the Depression Code is presented with a bottlenecked cookie jar. According to the need to get, that individual may stick a hand in the jar and grab as many cookies as that hand can hold. However, upon attempting to extract the cookies the narrow jar opening frustratingly imprisons the hand, rendering it powerless.

Now look for the conviction-related dilemma.

In the attempt to get, there was a loss of control. Conversely, to regain control, some of the getting has to be sacrificed. At this point, there's a conflict between the need to get and the need to be in control. It's a frustrating situation as any solution imperils life style conviction integrity.

Dysthymic individuals can face the Cookie Jar Conflict in countless ways.

For example,
- Get safety (or save money, not waste time) by not participating in some event, but lose control of how others perceive and judge them.
- Control their environment and the people in it, but not getting genuinely close relationships that require fairness, vulnerability, and periodically letting someone else have control.

Stoplight Conflict

Envision a group of burly thugs who approach a car at a stoplight and begin viciously attacking the vehicle. Now consider the driver's options.

The driver may take control of the situation by running the red light and driving away and thereby end the attack. But that would break the law and violate the need to be good, perfect, right conviction.

On the other hand, obeying the stoplight would enable the person to be good, but allow the thugs to overpower the driver, which would breach the need to be in control.

Either act violates a life style conviction.

Now consider how this might show up in more likely situations.
- Dysthymic persons may want to be the good person who always donates time and attention to others, but in doing so loses control of free time and personal pursuits.
- They control their time and effort (e.g., avoid seeing their friends in a play, not go to birthday parties or funerals, not volunteer), but risk "being bad" to friends, family, and the less fortunate.

The Candy Bar, Cookie Jar, and Stoplight conflicts concisely illustrate how the Depression Code convictions can be incompatible and relate to symptom development.

Perhaps it's easier to view the Depression Code as a family of convictions…complete with a troublesome tendency for sibling rivalry. But, there's more to dysthymia than infighting among convictions.

Given the social aspect of symptoms, and how each individual continually swims among a sea of other people, think about another way in which an individual's life style convictions can be associated with depression.

Common Sense vs. Private Logic

Most people's private logic easily fits within the framework of common sense. Some private sense can be more advanced, realistic, and healthier than common knowledge, and may benefit the individual as well as humanity.

Yet, symptoms can arise when a person's private logic is at odds with common sense, unrealistic, inflexible, and detrimental.

Like a business team member who stubbornly executes one plan when the rest of the group works on another, dysthymic individuals' private logic can drastically diverge from common knowledge and risk rejection, ridicule, failure, and abandonment.

While those consequences may seem like a sufficient deterrent, there's a notable catch.

People can believe their private logic is more accurate and useful in goal achievement than common knowledge. Now consider what happens when people's private logic conflicts with common sense.

Individuals seek to act in accord with their private logic.

While this seems like an impasse on par with a U.S. government shutdown, there's a common "solution" to this dilemma.

To not imperil social acceptance, neurotic individuals often attempt to abide by real (social) and imaginary (private logic) obligations.[245]

Now think about what this might look like.

Those with persistent depressive disorder may try to fit in with others by going along with social conventions, *all while living by their private logic*.

For example, imagine that a dysthymic individual secretly hates a co-worker that everyone else adores. When that beloved co-worker's birthday arrives, the dysthymic person signs the birthday card (perhaps with a suspiciously terse, obligatory acknowledgement or doesn't add any gift money to what's in the envelope). In this way, there's an acceptable display of going along with the group (i.e., not being the only one who didn't sign the card). But, as no one was tracking who put what money in the envelope, the stealthy act of rebellion enables adherence to private logic.

Given that dysthymic individuals can be intricately aware of what's good, perfect, right, they usually know the socially appropriate things to say and how to act in front of others to not jeopardize acceptance, status, or their way of life, but ultimately follow their private logic.

Note, this "double agent" type of behavior is common among neurotic and non-neurotic individuals alike.

For example, to avoid violating conventions and social fallout people may:

- Make impressive statements about some charity, but never volunteer.
- Indulge someone by laughing at a dull joke (while cringing internally).
- Tell everyone about the benefits of marriage, but cheat on their spouse.
- Listen to a boring story for the third time (as they cognitively vacation somewhere far more interesting).
- Tell others about how wonderful it is to be a parent, but regularly yell at their kids in private, or not acknowledge what they do well.
- When around others who regularly hug when greeting, they bristle at the socially-enforced rule and may hug briefly, stiffly, or half-heartedly.
- Visit cranky, demanding relatives out of "respect", but while there, secretly plan the rest of their day and count down the time remaining before they can leave without looking bad.

People can protect acceptance and status while obeying their private logic.

Knowing that, think about why anxiety is depression's frequent travel companion.

Anxiety increases when neurotic logic has more frequent or in-depth discord with the environment and the risk of failure escalates.[246]

Okay, but consider why that is.

Remaining vigilant, guarded, and able to act quickly is helpful when quietly maintaining a private logic that conflicts with common sense.

Next, given the need to be good, perfect, right you can also recognize that private logic can prompt incredibly kind, thoughtful acts.

Now consider the following examples to see how that can be related to developing depression.

- People let their estranged spouse keep all of the money and possessions in a divorce, because they want to be good to the person and don't want to look selfish or bad, but then experience devastating financial distress.
- They help out friends or family (e.g., moving, painting, yardwork) to the point of injury and spend several days unable to work, feeling conflicted, or upset about their self-inflicted injury and related loss of income.
- They worry more about other people's problems than those whose concern it is, and may consequently lose sleep, experience a decreased appetite, be distracted, or become despondent and gloomy.
- Regularly aid an ungrateful friend who never returns the effort.
- Even when struggling to pay their bills, they generously and empathetically give money to an unmotivated child with an unhealthy sense of entitlement or a co-worker who spends it frivolously.

Dysthymic persons can have private logic that's in accord with applauded social conventions and benefits others, yet they can develop depression when living by private logic that's beyond what's healthy.

So, why would individuals violate or overdo socially-acceptable guidelines?

Neurotic persons can take an individualized goal of achievement and measure their accomplishments from within their private logic.[247]

As they believe their private sense is more realistic and practical than the common sense, they may readily dismiss the social standards as either inaccurate or insufficient, then disobey or overextend them.

Also consider this symptom-related incompatibility.

Private logic that differs from common sense creates two different yardsticks that leads to either over- or under-valuing other people or themselves.

For example, those with neurotic depression may view others' possessions, skills, accomplishments, etc. against common sense standards and readily accept and praise them; *yet judge themselves by their private logic* and have a self-image of being inferior even when they have the same or greater abilities and attainments. This can perpetuate their depressive symptoms.

Dysthymic persons may measure themselves differently from what's commonly employed, as well as assess others by some index that isn't a shared standard. For instance, they may raise the bar of success for their children (actions, education, generosity, flexibility, etc.) that's far above what's commonplace. This double standard can lead to depressive symptoms when people realize that their children aren't living up to that uncommon measure.

When asked how they're feeling, dysthymic individuals may keep two sets of books: How they feel and What they tell others. They may do this as a way to not upset or concern them, or to avoid additional questions they deem as uncomfortably intrusive, as an example. But, not being genuinely vulnerable with others can lead to a depressing sense of loneliness.

People may feel overwhelmed, inferior, and develop depression when their life style convictions are unhealthy, inflexible, inadequate, and incompatible – either with each other or with common sense.

But, the mystery of how persistent depression can rise, survive, and thrive still isn't fully resolved.

Cliffhanger

Why people experience psychological symptoms and how to treat them is an intriguing mystery. Depression regularly lessens people's quality of life and can appear indecipherable and irresolvable. Nevertheless, throughout history assorted attempts have been made to understand and remedy symptoms.

Early cause and effect mindsets could be toxically incorrect. Fortunately, over the course of centuries flawed perspectives and solutions have mostly faded away, allowing more accurate, efficient, and effective assessment and treatment to remain.

Presently, the cause and effect mindset is productively employed routinely to understand and treat depression. Psychiatrists and psychologists strive to identify telltale causal agents that sabotage people's cognitive, physical, social, and emotional functioning. Various interventions address the perceived causes.

There are many ways in which people can become depressed. For instance, they may feel inferior, be ridiculed or rejected, be frustrated that others do not think and act as expected, feel out of sync with their surrounding community, fail at achieving their desires, not live up to their conception of a "Real" man or a "Real" woman, experience conflicts among their life style convictions, be painfully uncertain about their goals in life, and so on.

However, the cause and effect mindset has some intriguing and revealing shortcomings (e.g., insight inconsistencies, overuse of memory for assessment, seeing people as passive, prompting a search for a symptom cause, laypersons' overreliance on a chemical imbalance origin, unverified assumptions of genetic issues, and resolute belief all symptoms are unassailably hardwired).

Fortunately, the growth model addresses limitations associated with a cause and effect mindset as well as advocates the adoption of healthy rules and goals to prevent or reduce psychological symptoms.

There are countless clues – clinical and otherwise – which have guided you in decoding and solving the mystery of how and why persistent depressive disorder rises, survives, and thrives.

In fact, there's been a medley of mysteries. Chapters presented questions, perspectives, and answers that propelled the narrative forward, providing a wider view of dysthymia and clues to a comprehensive solution.

People are goal-oriented beings, who ultimately strive to survive and thrive. They're active individuals who desire to grow from an inferior state towards competency in the life tasks. However, they cannot achieve this individually.

Humans are social beings. They're undeniably reliant on lifelong socialization, continually seeking and requiring social connection and exchange. Through communication and consensus people create common sense; the generally-accepted and advocated rules, knowledge, and goals.

The social aspect of a person's symptoms requires detailed investigation.

People supremely value and desire acceptance, interaction, friendship, respect, and love. This is underscored by the sadness and dread associated with ridicule, rejection, failure, and abandonment. An individual's symptoms are often understood by viewing them against a social background.

While diverse goals and logic can advance humanity and provide a spectrum of perspectives to assist in individual and group survival and thriving, people can problematically clash.

Life style convictions compose each individual's personality, the software that powerfully shapes perception, action, emotion, and desires. Dysthymic depression can rise when a person's life style convictions are discordant with common sense, unrealistic, incorrect, inflexible, detrimental, overly applied, inappropriately used, and conflict with one another. Also, those with dysthymia hold an identifying constellation of convictions – the Depression Code – that makes them more likely to develop symptoms.

What life style convictions people select and cultivate, combined with their fictional finalism (their created and vague, but influential goal), can be not only related to symptom development, but also rather revealing.

The growth model and all of the associated elements provides an additional, complementary perspective that can be used in conjunction with the cause and effect mindset to productively assess and treat dysthymia.

But there's much more to learn.

In fact, there have been clues placed along the way that hint at an inescapable and pivotal question, the answer to which is required to fully decode persistent depression.

For example, recollect how biological symptoms demonstrate a genetically-programmed orchestration of assorted bodily systems, actions, and chemicals, which occurs automatically, without cognition, to assist survival and thriving.

Indeed, bodily symptoms can serve various purposes: provide feedback that identifies a threat, initiate action in accord with people's preferences, shape future cognition and behavior, and have a social factor as they can influence others' thoughts and movement which may benefit the symptomatic person.

With that in mind, what if psychological symptoms aren't necessarily the result of some cause, but personality-based means to achieve goals in the life tasks to help people survive and thrive?

To be continued…

References

[1] Shulman, B. H. (1985). Cognitive therapy and the individual psychology of Alfred Adler. In M. J. Mahoney & A. Freeman (Eds.), *Cognition and psychotherapy* (pp. 243-258). New York: Plenum.

[2] Mosak, H. H. (1973). *Alfred Adler: His influence on psychology today*. Park Ridge, NJ: Noyes Press.

Maniacci, M. P. (2002). The DSM and Individual Psychology: A general comparison. *Journal of Individual Psychology, 58*(4), 356-362.

Ellenberger, H. F. (1970). The discovery of the unconscious: The history and evolution of Dynamic psychiatry. New York: Basic Books.

Stein, H. T. (2008). Adler's legacy: Past, present, and future. *Journal of Individual Psychology, 64*(1), 4-20.

[3] Mosak, H. H. (1973). *Alfred Adler: His influence on psychology today*. Park Ridge, NJ: Noyes Press.

Maniacci, M. P. (2002). The DSM and Individual Psychology: A general comparison. *Journal of Individual Psychology, 58*(4), 356-362.

Lombardi, D. N., Florentino, M. C., and Lombardi, A. J. (1998). *Perfectionism and abnormal behavior. Journal of Individual Psychology, 54*(1), 61-71.

Ellenberger, H. F. (1970). The discovery of the unconscious: The history and evolution of Dynamic psychiatry. New York: Basic Books.

Stein, H. T. (2008). Adler's legacy: Past, present, and future. *Journal of Individual Psychology, 64*(1), 4-20.

[4] Adler, A. (1996b). What is neurosis? *Individual Psychology, 52*(4), 318-333. (Original work published 1935)

Mozdzierz, G. J. (1996). Adler's "What is neurosis?": Clinical and predictive revelations from the past. *Individual Psychology: The Journal of Adlerian Theory, Research and Practice, 52*(4), 342-350.

[5] Kessler, R. C., Chiu, W. T., Demler, O., and Walters, E. E. (2005). Prevalence, severity, and comorbidity of twelve-month DSM-IV disorders in the National Comorbidity Survey Replication (NCS-R). *Archives of General Psychiatry, 62*(6), 617-627.

Kessler, R. C., Berglund, P., Demler, O., Jin, R., and Walters, E. E. (2005). Lifetime prevalence and age-of-onset distributions of DSM-IV disorders in the National Comorbidity Survey replication. *Archives of General Psychiatry, 62*(6), 593-602.

Kessler, R. C., Berglund, P., Demler, O., Jin, R., Koretz, D., et al. (2003). The epidemiology of major depressive disorder: Results from the National Comorbidity Survey Replication (NCS-R). *Journal of the American Medical Association, 18, 289*(23), 3095-3105.

Weissman, M. M., Bland, R. C., Canino, G. J., Faravelli, C., Greenwald, S., et al. (1996). Cross-national epidemiology of major depression and bipolar disorder. *Journal of the American Medical Association, 276*, 293-299.

Sadock, B. J. and Sadock, V. A. (2007). *Kaplan and Sadock's synopsis of psychiatry: Behavioral sciences/clinical psychiatry. (10th ed.).* Philadelphia, PA: Wolter Kluwer/Lippincott Williams and Wilkins.

[6] World Health Organization. (2012, October 10). DEPRESSION: *A global crisis: World Mental Health Day.* Retrieved from http://www.who.int/mental_health/management/depression/wfmh_paper_depression_wmhd_2012.pdf

Dotson, V. M. (2017). Variability in depression: What have we been missing? *The American Journal of Geriatric Psychiatry, 25*(1), 23-24.

World Health Organization (2017, March 30). "Depression: Let's talk" says WHO, as depression tops list of causes of ill health. Retrieved from http://www.who.int/mediacentre/news/releases/2017/world-health-day/en/

[7] Singh-Manoux, A., Akbaraly, T. N., Marmot, M., Melchior, M., Ankri, J., et. al. (2010). Persistent depressive symptoms and cognitive function in late midlife: The Whitehall II study. *The Journal of Clinical Psychiatry, 71*(10), 1379-1385.

Dotson, V. M. (2017). Variability in depression: What have we been missing? *The American Journal of Geriatric Psychiatry, 25*(1), 23-24.

[8] Blehar, M. D. and Oren, D. A. (1997). Gender differences in depression. *Medscape Women's Health, 2*(2), 3. Revised from: Women's increased vulnerability to mood disorders: Integrating psychobiology and epidemiology. *Depression* (1995). 3, 3-12.

[9] Addis, M. E. and Mahalik, J. R. (2003). Men, masculinity, and the contexts of help seeking. *American Psychologist, 58*(1), 5-14.

Galdas, P., Darwin, Z., Kidd, L., Blickem, C., McPherson, K., et. al. (2014). The accessibility and acceptability of self-management support interventions for men with long term conditions: A systematic review and meta-synthesis of qualitative studies. *Bmc Public Health, 14*(1), 1230-1250.

Galdas, P., Fell, J., Bower, P., Kidd, L., Blickem, C., et. al. (2015). The effectiveness of self-management support interventions for men with long-term conditions: A systematic review and meta-analysis. *BMJ Open, 5*(3), e006620. http://doi.org/10.1136/bmjopen-2014-006620

Centers for Disease Control and Prevention (2001). Utilization of ambulatory medical care by women: United States, 1997-98, 13(149), 1-46.

[10] American Psychiatric Association (2000). *Diagnostic and statistical manual of mental disorders: DSM-IV-TR.* Washington, DC: American Psychiatric Association.

American Psychiatric Association (2013). *Diagnostic and statistical manual of mental disorders: DSM-V.* Washington, DC: American Psychiatric Association.

World Health Organization. (2018). The ICD-10 classification of mental and behavioural disorders: Clinical descriptions and diagnostic guidelines. Geneva: World Health Organization.

[11] World Health Organization. (2018). The ICD-10 classification of mental and behavioural disorders: Clinical descriptions and diagnostic guidelines. Geneva: World Health Organization.

American Psychiatric Association (2013). *Diagnostic and statistical manual of mental disorders: DSM-V*. Washington, DC: American Psychiatric Association.

[12] American Psychiatric Association (2013). *Diagnostic and statistical manual of mental disorders: DSM-V*. Washington, DC: American Psychiatric Association.

World Health Organization. (2018). The ICD-10 classification of mental and behavioural disorders: Clinical descriptions and diagnostic guidelines. Geneva: World Health Organization.

[13] McCullough, J. P. (2000). Treatment for chronic depression: Cognitive behavioral analysis system of psychotherapy (CBASP). New York: Guilford Press.

[14] Mosak, H. H. and Phillips, K. (1980). *Demons, germs and values*. Chicago: Alfred Adler Institute.

[15] psychology. (n.d.). Online Etymology Dictionary. Retrieved from Dictionary.com website: http://dictionary.reference.com/browse/psychology

[16] Mosak, H. H. and Phillips, K. (1980). *Demons, germs and values*. Chicago: Alfred Adler Institute.

Gross, C. C. (2003). Trepanation from the Palaeolithic to the Internet. In R. Arnott, S. Finger, & C. U. M. Smith (Eds.), with the support of B. Lichterman and R. Breitweiser. *Trepanation: Discovery, History, Theory* (pp. 307-322). Lisse, Netherlands: Swets and Zeitlinger Publishers.

[17] Mosak, H. H. and Phillips, K. (1980). *Demons, germs and values*. Chicago: Alfred Adler Institute.

[18] Mosak, H. H. and Phillips, K. (1980). *Demons, germs and values*. Chicago: Alfred Adler Institute.

[19] Mosak, H. H. and Phillips, K. (1980). *Demons, germs and values*. Chicago: Alfred Adler Institute.

[20] Mosak, H. H. and Phillips, K. (1980). *Demons, germs and values*. Chicago: Alfred Adler Institute.

[21] Vasile, R. G. (2001). Medical treatment of depression. In J. L. Jacobson & A. M. Jacobson (Eds.), *Psychiatric Secrets* (2nd ed., pp. 241-251). Philadelphia, PA: Hanley and Belfus.

[22] Vasile, R. G. (2001). Medical treatment of depression. In J. L. Jacobson & A. M. Jacobson (Eds.), *Psychiatric Secrets* (2nd ed., pp. 241-251). Philadelphia, PA: Hanley and Belfus.

Keller, M. B., McCullough, J. P., Klein, D. N., Arnow, B., Dunner, D. L. et al. (2000). A comparison of Nefazodone, the Cognitive Behavioral-Analysis system of psychotherapy, and their combination for the treatment of chronic depression. *New England Journal of Medicine, 342*(20), 1462-1470.

Weiner, I. B. and Craighead, W. E. (Eds.). (2010). *The Corsini encyclopedia of psychology: Fourth Edition, Volume 2.* Hoboken, NJ: John Wiley & Sons.

[23] Soling, C. and Kabillio, E. (Producers), and Kabillio, E. (Director). (1998). *A hole in the head.* [Motion picture]. United States of America: Spectacle Films.

[24] Interlandi, J. (2016, February 5). New estimate boosts the human brain's memory capacity 10-fold. *Scientific American.* Retrieved from https://www.scientificamerican.com/article/new-estimate-boosts-the-human-brain-s-memory-capacity-10-fold/

Steck, A. and Steck, B. (2015). *Brain and mind: Subjective experience and scientific objectivity.* New York: Springer International Publishing.

Denes, G. (2016). Neural plasticity across the lifespan: How the brain can change. London: Routledge.

Byrne, J. H., Fioravante, D., and Antzoulatos, E. G. (2006). Cellular and molecular mechanisms of associative and nonassociative learning. In M. E. Selzer, S. Clarke, L. G. Cohen, P. W. Duncan, and F. H. Gage (Eds.), *Textbook of neural repair and rehabilitation, Volume I* (pp. 79-94). New York: Cambridge University Press.

Ramirez-Amaya, V. (2007). Molecular mechanisms of synaptic plasticity underlying long-term memory formation. In F. Bermudez-Rattoni (Ed.), Neural plasticity and memory: From genes to brain imaging (pp. 47-66). New York: CRC Press.

Akers, K. G., Martinez-Canabal, A., Restivo, L., Yiu, A. P., De, C. A., et. al. (2014). Hippocampal neurogenesis regulates forgetting during adulthood and infancy. *Science, 344*(6184), 598-602.

Shen, H. (2014, May 8). New brain cells erase old memories. *Nature.* Retrieved from https://www.nature.com/news/new-brain-cells-erase-old-memories-1.15186#/b1

Bartol, T. M., Bromer, C., Kinney, J., Chirillo, M. A., Bourne, J. N., et. al. (2015). Nanoconnectomic upper bound on the variability of synaptic plasticity. *eLife, 4.*

Cowan, N. (2016). *Working memory capacity: Classic edition.* New York: Routledge.

Underwood, E. (2014, May 8). How the brain deletes old memories. *Science magazine.* Retrieved from http://www.sciencemag.org/news/2014/05/how-brain-deletes-old-memories

[25] Schacter, D. L. (2001). The seven sins of memory: How the mind forgets and remembers. Boston, MA: Houghton Mifflin.

Schacter, D. L., Chiao, J. Y., and Mitchell, J. P. (2003). *The seven sins of memory: Implications for self.* Annals of the New York Academy of Sciences, 1001, 226-239.

Ebbinghaus, H. (1885/1964). *Memory: A contribution to experimental psychology*, (H. A. Ruger and C. E. Bussenius, Trans.). New York: Dover Publications.

[26] Schacter, D. L. (1996). *Searching for memory.* New York: Basic Books.

Bartlett, F. C. (1932). Remembering: A study in experimental and social psychology. New York: Macmillan.

Mosak, H. H. and Di Pietro, R. (2005). *Early recollections: Interpretative method and application.* New York: Routledge.

Loftus E. F., Coan J., and Pickrell, J. E. (1996). Manufacturing false memories using bits of reality. In L. M. Reder, (Ed.), *Implicit memory and metacognition* (pp. 195-220). Mahwah, NJ: Lawrence Erlbaum.

Loftus, E. F. (1997). Creating false memories. *Scientific American, 277*, 70-75.

[27] Strange, D., Garry, M., Bernstein, D. M., and Lindsay, D. S. (2011). Photographs cause false memories for the news. *Acta Psychologica, 136*(1), 90-94.

Garry, M. and Gerrie, M. P. (2005). When photographs create false memories. *Current Directions in Psychological Science, 14*(6), 321-325.

[28] Carroll, R. T. (2003). The skeptic's dictionary: A collection of strange beliefs, amusing deceptions, and dangerous delusions. Hoboken, NJ: John Wiley & Sons.

[29] Seligman, M. E. P. (1971). Phobias and preparedness. *Behavior Therapy, 2,* 307-320.

Leonard, D. C. (2002). *Learning theories: A to z.* Westport, CT: Greenwood Press.

[30] Aronson, E. (1999). *The social animal.* New York: Worth.

Nickerson, R. S. (1998). Confirmation bias: A ubiquitous phenomenon in many guises. *Review of General Psychology, 2*(2), 175-220.

[31] Bartlett, F. C. (1932). Remembering: A study in experimental and social psychology. New York: Macmillan.

Mosak, H. H. and Di Pietro, R. (2005). *Early recollections: Interpretative method and application.* New York: Routledge.

[32] Mosak, H. H. and Di Pietro, R. (2005). *Early recollections: Interpretative method and application.* New York: Routledge.

[33] Moscovitch, M., Chein, J. M., Talmi, D., and Cohn, M. (2007). Learning and memory. In B. J. Baars & N. M. Gage (Eds.), *Cognition, brain, and consciousness: Introduction to cognitive neuroscience* (pp. 255-291). Amsterdam, The Netherlands: Elsevier.

Mosak, H. H. and Di Pietro, R. (2005). *Early recollections: Interpretative method and application.* New York: Routledge.

[34] Loftus E. F., Coan J., and Pickrell, J. E. (1996). Manufacturing false memories using bits of reality. In L. M. Reder, (Ed.), *Implicit memory and metacognition* (pp. 195-220). Mahwah, NJ: Lawrence Erlbaum.

Loftus, E. F. (1997). Creating false memories. *Scientific American, 277*, 70-75.

Lommen, M. J., Engelhard, I. M., and van den Hout, M. A. (2013). Susceptibility to long-term misinformation effect outside of the laboratory. *European Journal of Psychotraumatology,* 4. Retrieved from http://www.eurojnlofpsychotraumatol.net/index.php/ejpt/article/download/19864/pdf_1

[35] Bartlett, F. C. (1932). Remembering: A study in experimental and social psychology. New York: Macmillan.

Mosak, H. H. and Di Pietro, R. (2005). *Early recollections: Interpretative method and application*. New York: Routledge.

[36] Patihis, L., Frenda, S. J., LePort, A. K. R., Petersen, N., Nichols, R. M., et al. (2013). False memories in highly superior autobiographical memory individuals. *Proceedings of the National Academy of Sciences, 110*(52), 20947-20952.

[37] Loftus, E. F. (1996). *Eyewitness Testimony*. Cambridge, MA: Harvard University Press.

Cutler, B. L. and Penrod, S. D. (1995). *Mistaken identification: The eyewitness, psychology, and the law*. Cambridge, MA: Harvard University Press.

[38] Mosak, H. H. and Di Pietro, R. (2005). *Early recollections: Interpretative method and application*. New York: Routledge.

[39] Mosak, H. H. and Di Pietro, R. (2005). *Early recollections: Interpretative method and application*. New York: Routledge.

Mosak, H. H. (1958). Early recollections as a projective technique. *Journal of Projective Techniques, 22*(3), 302–311.

[40] Sompayrac, L. M. (2015). *How the immune system works*. Chichester, West Sussex: Wiley-Blackwell.

Patel, P. and Chatterjee, S. (2017). Innate and adaptive immunity: Barriers and receptor-based recognition. In S. Chatterjee, W. Jungraithmayr, and. Bagchi (Eds.), *Immunity and inflammation in health and disease* (pp. 3-14). London: Elsevier.

Lyons, I. (2011). *Lecture notes*. Chichester, West Sussex, UK: Wiley-Blackwell.

[41] Mosak, H. H. and Di Pietro, R. (2005). *Early recollections: Interpretative method and application*. New York: Routledge.

[42] World Health Organization. (2018). The ICD-10 classification of mental and behavioural disorders: Clinical descriptions and diagnostic guidelines. Geneva: World Health Organization.

Goldstein, I. (2006). Women's sexual function and dysfunction: Study, diagnosis and treatment. London: Taylor & Francis.

Fagan, P. J. (2004). *Sexual disorders: Perspectives on diagnosis and treatment*. Baltimore, MD: Johns Hopkins University Press.

[43] MacHale, S. M., O'Rourke, S. J., Wardlaw, J. M., and Dennis, M. S. (1998). Depression and its relation to lesion location after stroke. *Journal of Neurology, Neurosurgery, and Psychiatry, 64*(3), 371-374.

Pringle, A. M., Taylor, R., and Whittle, I. R. (1999). Anxiety and depression in patients with an intracranial neoplasm before and after tumour surgery. *British Journal of Neurosurgery, 13*(1), 46-51.

Tresch, D. D. (1998). Management of the older patient with acute myocardial infarction: Difference in clinical presentation between older and younger patients. *Journal of the American Geriatrics Society, 46*(9), 1157-1162.

Tresch, D. D. (1987). Atypical presentations of cardiovascular disorders in the elderly. *Geriatrics, 42*(10), 31-36.

Ghaemmaghami, C. A. and Brady, W. J. (2001). Pitfalls in the emergency department diagnosis of acute myocardial infarction. *Emergency Medicine Clinics of North America, 19*(2), 351-369.

Jiang, H.-K. and Chang, D.-M. (1999). Non-steroidal anti-inflammatory drugs with adverse psychiatric reactions: Five case reports. *Clinical Rheumatology, 18*(4), 339-345.

Sorgi, P., Ratey, J., Knoedler, D., Arnold, W., and Cole, L. (1992). Depression during treatment with beta-blockers: Results from a double-blind placebo-controlled study. *The Journal of Neuropsychiatry and Clinical Neurosciences, 4*(2), 187-189.

Gerstman, B. B., Jolson, H. M., Bauer, M., Cho, P., Livingston, J. M., and Platt, R. (1996). The incidence of depression in new users of beta-blockers and selected antihypertensives. *Journal of Clinical Epidemiology, 49*(7), 809-815.

Ried, L. D., McFarland, B. H., Johnson, R. E., and Brody, K. K. (1998). Beta-blockers and depression: The more the murkier?. *The Annals of Pharmacotherapy, 32*(6), 699-708.

Ayd, F. J. (2000). *Lexicon of psychiatry, neurology, and the neurosciences.* Philadelphia, PA: Lippincott Williams & Wilkins.

Hutto B. (1998). Subtle psychiatric presentations of endocrine diseases. *Psychiatric Clinics of North America, 21*(4), 905-916.

Talbot-Stern, J. K., Green, T., and Royle, T. J. (2000). Psychiatric manifestations of systemic illness. *Emergency Medicine Clinics of North America, 18*(2), 199-209.

Bierer, M. F. and Bierer, B. E. (1994). Psychiatric symptoms of medical illness and drug toxicity. In S. E. Hyman & G. E. Tesar, (Eds.), *Manual of Psychiatric Emergencies*, 3rd edition (214-245). Boston, MS: Little, Brown, and Co.

Tardiff K. (1998). Unusual diagnoses among violent patients. *Psychiatric Clinics of North America, 21*(3), 567-576.

Fallon, B. A., Kochevar, J. M., Gaito, A., and Nields, J. A. (1998). The underdiagnosis of neuropsychiatric Lyme disease in children and adults. *Psychiatric Clinics of North America, 21*(3), 693-703.

Levin, B. E. and Katzen, H. L. (2005). Early cognitive changes and nondementing behavioral abnormalities in Parkinson's disease. *Advances in Neurology, 96*, 84–94.

[44] apophenia (n.d.). *The Skeptic's Dictionary.* Retrieved from http://www.skepdic.com/apophenia.html

Shermer, M. (December, 2008). Paternity: Finding meaningful patterns in meaningless noise. *Scientific American.* Retrieved from http://www.scientificamerican.com/article.cfm?id=patternicity-finding-meaningful-patterns

Shermer, M. (2011). *The believing brain: From ghosts and gods to politics and conspiracies--how we construct beliefs and reinforce them as truths*. New York: Times Books.

[45] Corsini, R. J. (1999). *Dictionary of psychology*. Philadelphia, PA: Brunner/Mazel, Taylor and Francis.

[46] Lambert, M-P and Herceg, Z. (2011). Mechanisms of epigenetic gene silencing. In H. I. Roach, F. Bronner, and R. O. C. Oreffo (Eds.). *Epigenetic aspects of chronic diseases* (pp. 41-54). London: Springer.

[47] Weiner, I. B., Stricker, G., and Widiger, T. A. (2012). *Handbook of psychology: Clinical psychology*. Hoboken, NJ: John Wiley & Sons.

Cowen, P., Harrison, P., and Burns, T. (2012). *Shorter Oxford textbook of psychiatry*. Oxford: Oxford University Press.

Smoller, J. W. Genetics of mood and anxiety disorders (2008). In J. W. Smoller, B. R. Sheidley, and M. T. Tsuang, (Eds.). *Psychiatric genetics: Applications in clinical practice* (pp. 131-176). Washington, DC: American Psychiatric Pub.

McGrath, P. J. and Miller, J. M. (2010). Co-occurring anxiety and depression: concepts, significance, and treatment implications. In H. B. Simpson, Y. Neria, R. Lewis-Fernández, and Schneier (Eds.). *Anxiety disorders: Theory, research, and clinical perspectives* (pp 90-102). Cambridge, UK: Cambridge University Press.

Lilienfeld, S. O., Sauvigné, K. C., Lynn, S. J., Cautin, R. L., Latzman, R. D., et. al. (2015). Fifty psychological and psychiatric terms to avoid: A list of inaccurate, misleading, misused, ambiguous, and logically confused words and phrases. *Frontiers in Psychology, 6*(1100), 1-15.

[48] Straub, R. O. (2017). *Health psychology: A biopsychosocial approach*. New York: Worth Publishers.

Andrews, L. W. (2010). *Encyclopedia of depression*. Santa Barbara, CA: Greenwood Press.

[49] Huttenlocher, P. R. (2009). *Neural Plasticity*. Cambridge, MA: Harvard University Press.

Shermer, M. (2015). Hardwired = permanent. In J. Brockman (Ed.). *This idea must die: Scientific ideas that are blocking progress* (pp. 100–103). New York: Harper.

Merzenich, M. M. (2013). *Soft-wired: How the new science of brain plasticity can change your life*. San Francisco, CA: Parnassus.

Lilienfeld, S. O., Sauvigné, K. C., Lynn, S. J., Cautin, R. L., Latzman, R. D., et. al. (2015). Fifty psychological and psychiatric terms to avoid: A list of inaccurate, misleading, misused, ambiguous, and logically confused words and phrases. *Frontiers in Psychology, 6*(1100), 1-15.

[50] Sadock, B. J., Sadock, B. J., Sadock, V. A., and Ruiz, P. (2015). *Kaplan & Sadock's synopsis of psychiatry: Behavioral sciences/clinical psychiatry*. Philadelphia, PA: Lippincott Williams & Wilkins.

Jung, Y-, J. and Namkoong, N. (2014) Alcohol: Intoxication and poisoning – diagnosis and treatment. In E. V. Sullivan & A. Pfefferbaum (Eds.), Alcohol and the nervous system (3rd Series, Volume 125, pp. 115- 121). New York: Elsevier.

First, M. B. and Tasman, C.-E. A. (2010). Clinical guide to the diagnosis and treatment of mental disorders, 2nd Edition. New York: John Wiley & Sons.

[51] Moldover, J. E., Goldberg, K. B., and Prout, M. F. (2004). Depression after traumatic brain injury: A review of evidence for clinical heterogeneity. *Neuropsychology Review, 14*(3),143–154

Parker, R. S. (1996). The spectrum of emotional distress and personality changes after minor head injury incurred in a motor vehicle accident. *Brain Injury, 10*(4), 287-302.

Jorge, R. E., Robinson, R. G., Moser, D., Tateno, A., Crespo-Facorro, B., et. al. (2004). Major depression following traumatic brain injury. *Archives of General Psychiatry, 61*(1), 42–50.

Aitken, L., Simpson, S., and Burns, A. (1999). Personality change in dementia. *International Psychogeriatrics, 11*(3), 263-271.

American Psychiatric Association (2000). *Diagnostic and statistical manual of mental disorders* (4th ed. Text Revision). Washington, DC: American Psychiatric Association

[52] Weyandt, L. L. (2005). *The physiological bases of cognitive and behavioral disorders*. Mahwah, NJ: Lawrence Erlbaum Associates.

Zahn-Waxler, C., Crick, N. R., Shirtcliff, E. A., and Woods, K. E. (2006). The origins and development of psychopathology in females and males. In D. Cicchetti & D. J. Cohen (Eds.), *Developmental Psychopathology, Volume one: Theory and method* (2nd ed., pp. 76-138). Hoboken, NJ: John Wiley & Sons.

Pihl, R. O. and Nantel-Vivier, N. (2005). Biological vulnerabilities to the development of psychopathology. In B. L. Hankin & J. R. Z. Abela (Eds.), *Development of psychopathology: A vulnerability-stress perspective*. Thousand Oaks, CA: Sage.

[53] Sadock, B. J., Sadock, B. J., Sadock, V. A., and Ruiz, P. (2015). *Kaplan & Sadock's synopsis of psychiatry: Behavioral sciences/clinical psychiatry*. Philadelphia, PA: Lippincott Williams & Wilkins.

Thase, M. E., Fava, M., Halbreich, U., Kocsis, J. H., Koran, L., et. al. (1996). A placebo-controlled, randomized clinical trial comparing sertraline and imipramine for the treatment of dysthymia. *Archives of General Psychiatry, 53*(9), 777-784.

Weiner, I. B. and Craighead, W. E. (Eds.) (2010). *The Corsini encyclopedia of psychology: Fourth Edition, Volume 2*. Hoboken, NJ: John Wiley & Sons.

Shirley, A. C. (2006). *Focus on serotonin uptake inhibitor research*. New York: Nova Science Publishers.

Pies, R. W. and Rogers, D. P. (2005). *Handbook of essential psychopharmacology.* Washington, DC: American Psychiatric Pub.

Sadock, B. J., Sadock, V. A., Ruiz, P., and Sadock, B. J. (2017). *Kaplan & Sadock's concise textbook of clinical psychiatry.* Philadelphia, PA: Lippincott Williams & Wilkins.

Vasile, R. G. (2001). Medical treatment of depression. In J. L. Jacobson & A. M. Jacobson (Eds.), *Psychiatric Secrets* (2nd ed., pp. 241-251). Philadelphia, PA: Hanley and Belfus.

Keller, M. B., McCullough, J. P., Klein, D. N., Arnow, B., Dunner, D. L. et al. (2000). A comparison of Nefazodone, the Cognitive Behavioral-Analysis system of psychotherapy, and their combination for the treatment of chronic depression. *New England Journal of Medicine, 342*(20), 1462-1470.

Weiner, I. B. and Craighead, W. E. (Eds.). (2010). *The Corsini encyclopedia of psychology: Fourth Edition, Volume 2.* Hoboken, NJ: John Wiley & Sons.

Kocsis, J. H. (2003). Pharmacotherapy for chronic depression. *Journal of Clinical Psychology, 59*(8), 885-892.

[54] Frosch, D. L., Krueger, P. M., Hornik, R. C., Cronholm, P. F., and Barg, F. K. (2007). Creating demand for prescription drugs: A content analysis of television direct-to-consumer advertising. *Annals of Family Medicine, 5*(1): 6-13.

Hollon, M. F. (1999). Direct-to-consumer marketing of prescription drugs: Creating consumer demand. *Journal of the American Medical Association, 281*(4), 382–384.

France, C. M., Lysaker, P. H., and Robinson, R. P. (2007). The "chemical imbalance" explanation for depression: Origins, lay endorsement, and clinical implications. *Professional Psychology Research and Practice, 38*(4), 411–420.

Deacon, B. J. and Baird, G. L. (2009). The chemical imbalance explanation of depression: Reducing blame at what cost? *Journal of Social and Clinical Psychology, 28*(4), 415–435.

Lilienfeld, S. O., Sauvigné, K. C., Lynn, S. J., Cautin, R. L., Latzman, R. D., et. al. (2015). Fifty psychological and psychiatric terms to avoid: A list of inaccurate, misleading, misused, ambiguous, and logically confused words and phrases. *Frontiers in Psychology, 6*(1100), 1-15.

[55] Young, S. N. (2007). How to increase serotonin in the human brain without drugs. *Journal of Psychiatry and Neuroscience, 32*(6), 394–399.

[56] Glenberg, A. M. and Andrzejewski, M. E. (2008). *Learning from data: An introduction to statistical reasoning (3rd ed.).* New York: Taylor and Francis.

[57] Salazar, L. F., DiClemente, R. J. and Crosby, R. A. (2006). Philosophy of science and theory construction. In R. A. Crosby, R. J. DiClemente, & L. F. Salazar (Eds.), *Research methods in health promotion* (p. 30). San Francisco, CA: Jossey-Bass.

[58] Fisher, L. (2011). *Crashes, crises, and calamities: How we can use science to read the early-warning signs.* New York: Basic Books.

Baardwijk, M. and Franses, P. H. (2010). *The hemline and the economy: Is there any match?*. Erasmus School of Economics. Retrieved from https://econpapers.repec.org/paper/emseureir/ 20147.htm

[59] Myers, D. G. and DeWall, C. N. (2018). *Psychology*. New York: Worth Publishers.

Myers, D. G. (2011). *Exploring psychology: Eighth edition in modules*. New York: Worth Publishers.

Fineburg, A. C., Blair-Broeker, C. T., Ernst, R. M., and Myers, D. G. (2013). *Thinking about psychology: The science of mind and behavior*. New York: Worth Publishers.

[60] Blows, W. T. (2005). The biological basis of nursing: Mental health. London: Routledge.

DeRubeis, R. J., Siegle, G. J., and Hollon, S. D. (2008). Cognitive therapy vs. medications for depression: Treatment outcomes and neural mechanisms. *Nature Reviews. Neuroscience, 9*(10), 788–796. http://doi.org/10.1038/nrn2345

Paquette, V., Lévesque, J., Mensour, B., Leroux, J. M., Beaudoin, G., et. al. (2003). "Change the mind and you change the brain": Effects of cognitive-behavioral therapy on the neural correlates of spider phobia. *Neuroimage, 18*(2), 401-409.

Arden, J. B. (2010). *Rewire your brain: Think your way to a better life*. Hoboken, NJ: John Wiley & Sons.

[61] Thase, M. E. and Lang, S. S. (2006). Beating the blues: New approaches to overcoming dysthymia and chronic mild depression. New York: Oxford University Press.

[62] Frosch, D. L., Krueger, P. M., Hornik, R. C., Cronholm, P. F., and Barg, F. K. (2007). Creating demand for prescription drugs: A content analysis of television direct-to-consumer advertising. *Annals of Family Medicine, 5*(1), 6-13.

Hollon, M. F. (1999). Direct-to-consumer marketing of prescription drugs: Creating consumer demand. *Journal of the American Medical Association, 281*(4), 382–384.

[63] Riederer, P., Sofic, E., Konradi, C., Kornhuber, J., Beckmann, H., et al. (1989). The role of dopamine in the control of neurobiological functions. In P. Riederer, (Ed.). *The role of brain dopamine* (pp. 1-18). New York: Springer-Verlag.

Thorner, M. O. and Vance, M. L. (1989). Clinical aspects of dopamine in the regulation of human anterior pituitary function. In P. Riederer, (Ed.). *The role of brain dopamine* (pp. 19-30). New York: Springer-Verlag.

Seo, D., Patrick, C. J., and Kennealy, P. J. (2008). Role of serotonin and dopamine system interactions in the neurobiology of impulsive aggression and its comorbidity with other clinical disorders. *Aggression and Violent Behavior, 13*(5), 383-395.

Müller, C. and Jacobs, B. L. (2010). *Handbook of the behavioral neurobiology of serotonin*. New York: Elsevier.

Iversen, L L., Iversen, S. D., Dunnett, S. B. and Borklund, A. (Eds.). (2010). *Dopamine handbook*. New York: Oxford University Press.

[64] Lilienfeld, S. O., Sauvigné, K. C., Lynn, S. J., Cautin, R. L., Latzman, R. D., et. al. (2015). Fifty psychological and psychiatric terms to avoid: A list of inaccurate, misleading, misused, ambiguous, and logically confused words and phrases. *Frontiers in Psychology, 6*(1100), 1-15.

[65] Lacasse, J. R., and Leo, J. (2005). Serotonin and depression: A disconnect between the advertisements and the scientific literature. *Public Library of Science, 2*(12), Med. 2:e392. doi: 10.1371/journal.pmed.0020392.

Leo, J., and Lacasse, J. R. (2008). The media and the chemical imbalance theory of depression. *Society, 45*(1), 35–45. doi: 10.1007/s12115-007-9047-3

Pies, R. (2011). Psychiatry's new brain-mind and the legend of the chemical imbalance. *Psychiatric Times*. Retrieved from http://www.psychiatrictimes.com/blogs/psychiatry-new-brain-mind-and-legend-chemical-imbalance

Lilienfeld, S. O., Sauvigné, K. C., Lynn, S. J., Cautin, R. L., Latzman, R. D., et. al. (2015). Fifty psychological and psychiatric terms to avoid: A list of inaccurate, misleading, misused, ambiguous, and logically confused words and phrases. *Frontiers in Psychology, 6*(1100), 1-15.

[66] apophenia (n.d.). *The Skeptic's Dictionary*. Retrieved from http://www.skepdic.com/apophenia.html

Shermer, M. (December, 2008). Paternity: Finding meaningful patterns in meaningless noise. *Scientific American*, Retrieved from http://www.scientificamerican.com/article.cfm?id= patternicity-finding-meaningful-patterns

Shermer, M. (2011). The believing brain: From ghosts and gods to politics and conspiracies--how we construct beliefs and reinforce them as truths. New York: Times Books.

[67] Vyse, S. A. (1997). *Believing in magic: The psychology of superstition*. New York: Oxford University Press.

[68] Mack, A. and Rock, I. (1998). *Inattentional blindness*. Cambridge, MA: MIT Press.

[69] Plowman, S. A. and Smith, D. L. (2008). *Exercise physiology for health, fitness, and performance (Reprinted 2nd ed.)*. Baltimore: Lippincott Williams and Wilkins.

West, B. J. and Griffin, L. A. (2004). *Biodynamics: Why the wirewalker doesn't fall*. Hoboken, NJ: John Wiley & Sons.

[70] Singer, A. J. and Clark, R. A. F. (2002). The biology of wound healing. In A. J. Singer & J. E. Hollancher (Eds.), *Lacerations and acute wounds: An evidence-based guide* (pp. 1-8). Philadelphia: F. A. Davis Company.

Trott, A. (2005). Wounds and lacerations: Emergency care and closure (3rd ed.). Philadelphia: Mosby.

[71] Chohan, N. D. (Ed.) (2008). *Nursing: Interpreting signs and symptoms*. Ambler, PA: Lippincott Williams and Wilkins.

Watson Genna, C. and Sandora, L. (2007). Normal sucking and swallowing. In C. Watson Genna. *Supporting sucking skills in breastfeeding infants* (pp. 1-42). Boston: Jones and Bartlett Publishers.

[72] Porth, C. M. (2011). *Essentials of pathophysiology: Concepts of altered health states*. Philadelphia: Lippincott Williams & Wilkins.

Plaford, G. R. (2009). *Sleep and learning: The magic that makes us healthy and smart*. Lanham, MD: Rowman & Littlefield Education.

[73] Naslund, E. and Hellstrom, P. M. (2007). Appetite signaling: From gut peptides and enteric nerves to brain. *Physiology and Behavior, 92*(1-2), 256-262.

Gray, P. O. (2002). Mechanisms of motivation, emotions, and sleep. In *Psychology (4th ed.)* (pp. 187-230). New York: Worth Publishers.

Woods, S. C. and Stricker, E. M. (2008). Food intake and metabolism. In L. R. Squire, D. Berg, F. Bloom, S. du Lac, & A. Ghosh (Eds.), *Fundamental neuroscience (3rd ed.)* (pp. 873-888). Burlington, MA: Academic Press.

[74] Taylor, P. N., Wolinsky, I., and Klimis, D. J. (1999). Water in exercise and sport. In J. A. Driskell, I. & Wolinsky (Eds.), *Macroelements, water, and electrolytes in sports nutrition* (pp. 93-108). Boca Raton, FL: CRC Press Inc.

Stricker, E. M. and Verbalis, J. G. (2008). Water intake and bodily fluids. In L. R. Squire, D. Berg, F. Bloom, S. du Lac, & A. Ghosh (Eds.), *Fundamental neuroscience (3rd ed.)* (pp. 889-930). Burlington, MA: Academic Press.

Rolls, B. J. (1993). Palatability and fluid intake. In B. M. Marriott (Ed.), *Fluid replacement and heat stress* (pp. 161-167). Washington, DC: Food and nutrition board, Institute of Medicine, National Academy of Sciences.

Woods, S. C. and Stricker, E. M. (2008). Food intake and metabolism. In L. R. Squire, D. Berg, F. Bloom, S. du Lac, & A. Ghosh (Eds.), *Fundamental neuroscience (3rd ed.)* (pp. 873-888). Burlington, MA: Academic Press.

[75] Miller, A. D. (1993). Neuroanatomy and physiology. In M. H. Sleisenger (Ed.). *The handbook of nausea and vomiting* (pp. 1-10). New York: Parthenon Publishing.

Horn, C. C. (2008). Why is the neurobiology of nausea and vomiting so important?. *Appetite, 50*(2-3), 430-434.

Rogers, D. F. (2004). Overview of airway mucus clearance. In B. K. Rubin, & C. P. van der Schans (Eds.), *Therapy for mucus-clearance disorders* (pp. 1-27). New York: Marcel Dekker, Inc.

Pus cell (2008, October 25). *Biology online*. Retrieved from http://www.biology-online.org/dictionary/Pus_cell

Sompayrac, L. M. (2008). *How the immune system works (3rd ed.)*. Malden, MA: Blackwell publishing.

Griffin, J., Arif, S., and Mufti, A. (2003). *Crash course: Immunology and haematology (2nd ed.)*. St. Louis, MO: C. V. Mosby.

Davidson, G. P. (1991). Viral diarrhea. In M. Gracey (Ed.), *Diarrhea*. Boca Raton, FL: Telford Press.

[76] Hellier, J. L. (2017). *The five senses and beyond: The encyclopedia of perception*. Santa Barbara, CA: Greenwood.

[77] National Center for Health Statistics. (2016). *Health, United States, 2015: With special feature on racial and ethnic health disparities*. Table 80: Selected prescription drug classes used in the past 30 days, by sex and age: United States, selected years 1988–1994 through 2009–2012. Hyattsville, MD.

Pratt, L.A., Brody, D.J., and Gu Q. (2017). *Antidepressant use in persons aged 12 and over: United States, 2011–2014*. NCHS data brief, no 283. Hyattsville, MD: National Center for Health Statistics.

[78] Pratt, L.A., Brody, D.J., and Gu Q. (2017). *Antidepressant use in persons aged 12 and over: United States, 2011–2014*. NCHS data brief, no 283. Hyattsville, MD: National Center for Health Statistics.

Pratt, L.A., Brody, D.J., and Gu Q. (2011). *Antidepressant use in persons aged 12 and over: United States, 2005–2008*. NCHS data brief, no 76. Hyattsville, MD: National Center for Health Statistics.

[79] OECD (2017), Pharmaceutical consumption. *Health at a Glance 2017: OECD Indicators*. OECD Publishing: Paris. http://dx.doi.org/10.1787/ health_glance-2017-70-en

Mars, B., Heron, J., Kessler, D., Davies, N. M., Martin, R. M., et. al. (2017). Influences on antidepressant prescribing trends in the UK: 1995–2011. *Social Psychiatry and Psychiatric Epidemiology*, *52*(2), 193–200. http://doi.org/ 10.1007/s00127-016-1306-4

Gould., S. and Friedman, L. F. (2016, February 4). Something startling is going on with antidepressant use around the world. *Business Insider*. Retrieved from http://www.businessinsider.com/countries-largest-antidepressant-drug-users-2016-2

[80] Shelton, R. C. (2006). The nature of the discontinuation syndrome associated with antidepressant drugs. *Journal of Clinical Psychiatry, 67*(4), 3–7.

Robinson, D. S. (2006). Antidepressant discontinuation syndrome. *Primary Psychiatry, 13*(10), 23-24.

Warner, C. H., Bobo, W., Warner, C., Reid, S., and Rachal, J. (2006). Antidepressant discontinuation syndrome. *American Family Physician, 74*(3), 449-456.

Zajecka, J., Tracy, K. A., and Mitchell, S. (1997). Discontinuation symptoms after treatment with serotonin reuptake inhibitors: a literature review. *Journal of Clinical Psychiatry, 58*(7), 291–297.

[81] Mosak, H. H. and Phillips, K. (1980). *Demons, germs and values*. Chicago: Alfred Adler Institute.

[81] Mosak, H. H. (1984). Adlerian psychotherapy. In R. Corsini & D. Wedding (Eds.), *Current psychotherapies* (3rd ed.) (pp. 56-107). Itasca, IL: F. E. Peacock Publishers.

Mosak, H. J. (1989). Adlerian psychotherapy, in R. J. Corsini & D. Wedding (Eds.), *Current psychotherapies* (4th ed.) (pp. 65-116), Itasca, IL: F. E. Peacock Publishers.

Oberst, U. E. and Stewart, A. E. (2012). Adlerian psychotherapy: An advanced approach to Individual Psychology. New York: Routledge.

Sperry, L. and Carlson, J. (2013). *Psychopathology and psychotherapy: From DSM-IV diagnosis to treatment*. Washington, DC: Taylor and Francis.

[82] Mosak, H. H. and Phillips, K. (1980). *Demons, germs and values*. Chicago: Alfred Adler Institute.

[83] Mosak, H. H. (1984). Adlerian psychotherapy. In R. Corsini & D. Wedding (Eds.), *Current psychotherapies* (3rd ed.) (pp. 56-107). Itasca, IL: F. E. Peacock Publishers.

Mosak, H. H. (1989). Adlerian psychotherapy. In R. J. Corsini & D. Wedding (Eds.), *Current psychotherapies* (4th ed.) (pp. 65-116), Itasca, IL: F. E. Peacock Publishers.

Sperry, L. and Carlson, J. (2013). *Psychopathology and psychotherapy: From DSM-IV diagnosis to treatment*. Washington, DC: Taylor and Francis.

Oberst, U. E. and Stewart, A. E. (2012). Adlerian psychotherapy: An advanced approach to Individual Psychology. New York: Routledge.

Obembe, S. (2012). Theories of treatment. In S. Obembe, (Ed.), *Practical skills and clinical management of alcoholism & drug addiction* (pp. 47-62). Chennai, Amsterdam: Elsevier.

[84] Mosak, H. H. and Maniacci, M. P. (1999). A primer of Adlerian psychology: The analytic-behavioral-cognitive psychology of Alfred Adler. Philadelphia, PA: Brunner/Mazel.

[85] Mosak, H. H. and Phillips, K. (1980). *Demons, germs and values*. Chicago: Alfred Adler Institute.

[86] Mosak, H. H. and Phillips, K. (1980). *Demons, germs and values*. Chicago: Alfred Adler Institute.

[87] Kricher, J. (1997). A neotropical companion: An introduction to the animals, plants, and ecosystems of the new world tropics (2nd ed.). Princeton, NJ: Princeton University Press.

Grossinger, R. (1995). *Planet medicine: Origins* (Rev. ed.). Berkeley, CA: North Atlantic Books.

Whyte, S.R. (1988). The power of medicines in East Africa. In S. van der Geest & S.R. Whyte (Eds.), *The context of medicines in developing countries. Studies in Pharmaceutical Anthropology* (pp. 217-233). Boston: Kluwer Academic Publishers.

Earthzine.org (2008, December 2). *Earthzine >> blog archive >> the threat to nature's medicine*. Retrieved from http://www.earthzine.org/2008/12/02/the-threat-to-natures-medicine/

The New York Times Company (1991, June 11). *Shamans and their lore may vanish with forests*. Retrieved from http://query.nytimes.com/gst/fullpage.html?res= 9D0CEEDA153AF932A25755C0A967958260&sec=health&spon=&pagewanted=all

Adams, M. (2003, November 8). Pharmaceutical companies steal from medicine men, all in the name of profit. Natural News.com Retrieved from http://www.naturalnews.com/000476.html

[88] Darwin, C. (1859). *On the origin of species by means of natural selection, or the preservation of favoured races in the struggle for life.* London: John Murray. Retrieved from http://darwin-online.org.uk/content/frameset?itemID=F373&viewtype=text&pageseq=1

Dawkins, R. (1989). *The selfish gene.* New York: Oxford University Press.

[89] Benyus, J. M. (2009). *Biomimicry: Innovation inspired by nature.* New York: Perennial.

Vincent, J. F., Bogatyreva, O. A., Bogatyrev, N. R., Bowyer, A., and Pahl, A. K. (2006). Biomimetics: Its practice and theory. *Journal of the Royal Society, Interface, 3*(9), 471-482.

Vox (2017, November 9). *The world is poorly designed. But copying nature helps.* Retrieved from https://www.youtube.com/watch?v=iMtXqTmfta0

[90] Dawkins, R. (1989). *The selfish gene.* New York: Oxford University Press.

[91] Niklas, K. J. (1997). The *evolutionary biology of plants.* Chicago: University of Chicago Press.

Pearson, L. C. (1995). *The diversity and evolution of plants.* Boca Raton, FL: CRC Press Inc.

[92] Takagi, S. (2002). Actin-based photo-orientation movement of chloroplasts in plant cells, *Journal of Experimental Biology 206*, 1963-1969. Retrieved from http://jeb.biologists.org/cgi/content/full/206/12/1963#REF31

[93] Kemp, T. S. (2005). *The origin and evolution of mammals.* New York: Oxford University Press.

[94] Caras, R. A. (1972). *Protective coloration and mimicry: Nature's camouflage.* Richmond, VA: Westover Pub. Co.

[95] Dawkins, R. (1989). *The selfish gene.* New York: Oxford University Press.

Hamilton, W. D. (1971). Geometry for the selfish herd. *Journal of Theoretical Biology, 31*, 295-311.

[96] Bergstrom, C. T. and Feldgarden, M. (2008). The ecology and evolution of antibiotic-resistant bacteria. In S. Stearns & J. Koella (Eds.), *Evolution in Health and Disease, Second Edition* (pp. 125-137). Oxford University Press. Retrieved from

http://octavia.zoology.washington.edu/publications/BergstromAndFeldgarden08.pdf

[97] Darwin, C. (1859). *On the origin of species by means of natural selection, or the preservation of favoured races in the struggle for life.* London: John Murray. Retrieved from http://darwin-online.org.uk/content/frameset?itemID=F373&viewtype=text&pageseq=1

Dawkins, R. (1989). *The selfish gene.* New York: Oxford University Press.

[98] Darwin, C. (1859). *On the origin of species by means of natural selection, or the preservation of favoured races in the struggle for life.* London: John Murray. Retrieved from http://darwin-online.org.uk/content/frameset?itemID=F373&viewtype=text&pageseq=1

[99] Nye, B. and Powell, C. S. (2014). *Undeniable: Evolution and the science of creation.* New York: St. Martin's Press.

It's okay to be smart. (2015, December 19). *Evolution Is Dumb - 12 Days of Evolution #6* [video file]. Retrieved from https://www.youtube.com/watch?v=7oLLzmNmrfo

Lehmberg, J. (2017, November 10). Nature up close: Survival of the "good enough". *CBS News.* Retrieved from https://www.cbsnews.com/news/nature-up-close-the-evolution-of-good-enough/

Coyne, J. (2015, December 26). *Why evolution is true.* Retrieved from https://whyevolutionistrue.wordpress.com/2015/12/26/the-12-days-of-evolution-6-the-imperfection-of-evolution/

[100] Williams, G. C. (1966). *Adaptation and natural selection.* Princeton: Princeton University Press. As cited in Dawkins, R. (1989). *The selfish gene.* New York: Oxford University Press.

Dawkins, R. (1989). *The selfish gene.* New York: Oxford University Press.

[101] Droma, T. S., McCullough, R. G., McCullough, R. E., Zhuang, J. G., Cymerman, A., Sun, S. F., et al. (1991). Increased vital and total lung capacities in Tibetan compared with Han residents of Lhasa (3658 m.). *American Journal of Physical Anthropology, 86*(3), 341-351.

[102] McEvoy, B., Beleza, S., and Shriver, M. D. (2006). The genetic architecture of normal variation in human pigmentation: An evolutionary perspective and model. *Human Molecular Genetics, 15*(2), R176-R181.

Darwin, C. (1871). *The decent of man and selection in relation to sex.* London: John Murray. Retrieved from http://darwin-online.org.uk/content/frameset?itemID=F937.1&viewtype=text&pageseq=1

[103] Fujimoto, A., Kimura, R., Ohashi, J., Omi, K., Yuliwulandari, R., et al. (2008). A scan for genetic determinants of human hair morphology: EDAR is associated with Asian hair thickness. *Human Molecular Genetics, 17*(6), 835–843.

[104] Beauchamp, J. P. (2016). Genetic evidence for natural selection in humans in the contemporary United States. *Proceedings of the National Academy of Sciences of the United States of America, 113*(28), 7774-7779.

Coyne, J. A. (2010). *Why evolution is true.* Oxford: Oxford University Press.

Zuk, M. (2014). Paleofantasy: What evolution really tells us about sex, diet, and how we live. New York: W.W. Norton & Company

Sadava, D. E. (2014). *Life: The science of biology.* Sunderland, MA: Sinauer Associates.

University of Chicago Medical Center. (2005, September 9). University of Chicago researchers find human brain still evolving. *ScienceDaily.* Retrieved from www.sciencedaily.com/ releases/2005/ 09/ 050909221043.htm

Balter, M. (2005). Are human brains still evolving? Brain genes show signs of selection. *Science, 309*(5741), 1662-1663.

Bromham, L. (2016). An introduction to molecular evolution and phylogenetics. Oxford: Oxford University Press.

Relethford, J. H. (2012). *Human population genetics.* Hoboken, NJ: Wiley-Blackwell.

Monosson, E. (2015). Unnatural selection: How we are changing life, gene by gene. Washington, DC: Island Press

Yashon, R. K. and Cummings, M. R. (2012). *Human genetics and society.* Belmont, CA: Brooks/Cole, Cengage Learning.

Tishkoff, S. (2015). Strength in small numbers. *Science, 349*(6254), 1282-1283.

Fumagalli, M., Moltke, I., Grarup, N., Racimo, F., Bjerregaard, P., et. al. (2015). Greenlandic Inuit show genetic signatures of diet and climate adaptation. *Science, 349*(6254), 1343-1347.

Byars, S. G., Ewbank, D., Govindaraju, D. R., and Stearns, S. C. (2010). Natural selection in a contemporary human population. *Proceedings of the National Academy of Sciences of the United States of America, 107*(1), 1787-1792.

[105] Dreikurs, R. and Soltz, V. (1990). *Children: The challenge.* New York: Plume.

[106] Cialdini, R. B. (1993). *Influence: The psychology of persuasion* (Rev. ed.). New York: Morrow.

Aronson, E. (1999). *The social animal* (8th ed.). New York: Freeman.

Sapolsky, R. M. (1998). Why zebras don't get ulcers: An updated guide to stress, stress-related diseases, and coping. New York: W. H. Freeman and Company.

[107] Clancy, B., Finlay, B. L., Darlington, R. B. and Anand, K. J. S. (2007). Extrapolating brain development from experimental species to humans. *Neurotoxicology 28*(5), 931-937.

[108] Adler, A. (1956). The Individual Psychology of Alfred Adler: A systematic presentation in selections from his writings. H. Ansbacher and R. Ansbacher (Eds.). New York: Basic Books.

[109] Adler, A. (1956). The Individual Psychology of Alfred Adler: A systematic presentation in selections from his writings. H. Ansbacher and R. Ansbacher (Eds.). New York: Basic Books.

[110] Adler, A. (1956). The Individual Psychology of Alfred Adler: A systematic presentation in selections from his writings. H. Ansbacher and R. Ansbacher (Eds.). New York: Basic Books.

[111] Dreikurs, R. and Mosak, H. H. (1967). The tasks of life II. The fourth life task. *Individual Psychologist, 4*(2), 51-56.

[112] Mosak, H. H. and Dreikurs, R. (1967). The life tasks III: The fifth life task. *Individual Psychologist, 5*, 16-22.

[113] Mosak, H. H. and Maniacci, M. P. (1999). A primer of Adlerian psychology: The analytic-behavioral-cognitive psychology of Alfred Adler. Philadelphia, PA: Brunner/Mazel.

[114] Adler, A. (1927). *Understanding human nature.* (W. B. Wolfe, Trans.). Garden City, NY: Garden City Publishing Company, Inc.

[115] White, R. W. (1959). Motivation reconsidered: The concept of competence. *Psychological Review, 66*(5), 297-333.

[116] Adler, A. (1956). The Individual Psychology of Alfred Adler: A systematic presentation in selections from his writings. H. Ansbacher and R. Ansbacher (Eds.). New York: Basic Books.

[117] Adler, A. (1929b). *The science of living.* New York: Greenberg, Publisher, Inc.

Adler, Alexandra. (1938). Guiding human misfits: A practical application of Individual Psychology. New York: Macmillan.

[118] Banerjee, J. C. (1994). *Encyclopaedic dictionary of psychological terms.* New Delhi: MD Publications.

[119] Banerjee, J. C. (1994). *Encyclopaedic dictionary of psychological terms.* New Delhi: MD Publications.

[120] Adler, A. (1929). *The practice and theory of Individual Psychology.* (P. Radin, Trans.). New York: Harcourt, Brace & Company, Inc.

[121] Goldfeld, S., Wise, P., and Zuckerman, B. (2002). The impact of maternal health on child health status and health service utilisation. *Journal of Paediatrics and Child Health, 38*(5): A8.

Debes, F., Budtz-Jørgensen, E., Weihe, P., White, R. F., and Grandjean, P. (2006). Impact of prenatal methylmercury exposure on neurobehavioral function at age 14 years. *Neurotoxicology and Teratology, 28*(5), Sep-Oct 2006, 536-547.

Burns, L., Mattick, R. P., and Wallace, C. (2008, June). Smoking patterns and outcomes in a population of pregnant women with other substance use disorders. *Nicotine and Tobacco Research, 10*(6), 969-974.

Shea, A. K., and Steiner, M. (2008). Cigarette smoking during pregnancy. *Nicotine and Tobacco Research, 10*(2), Feb 2008, 267-278.

Salihu, H. M., Sharma, P. P., Getahun, D., Hedayatzadeh, M., Peters, S., Kirby, R. S., et al. (2008, January). Prenatal tobacco use and risk of stillbirth: A case-control and bidirectional case-crossover study. *Nicotine and Tobacco Research, 10*(1), 159-166.

Cornelius, M. D., Goldschmidt, L., DeGenna, N., and Day, N. L. (2007, July). Smoking during teenage pregnancies: Effects on behavioral problems in offspring. *Nicotine and Tobacco Research, 9*(7), 739-750.

Yumoto, C., Jacobson, S. W., and Jacobson, J. L. (2008, November-December). Fetal substance exposure and cumulative environmental risk in an African American cohort. *Child Development, 79*(6), 1761-1776.

Vaurio, L., Riley, E. P., and Mattson, S. N. (2008, January). Differences in executive functioning in children with heavy prenatal alcohol exposure or attention-deficit/hyperactivity disorder. *Journal of the International Neuropsychological Society, 14*(1), 119-129.

Autti-Ramo, I., Autti, T., Korkman, M., Kettunen, S., Salonen, O., and Valanne, L. (2002, February). MRI findings in children with school problems who had been exposed prenatally to alcohol. *Developmental Medicine and Child Neurology, 44*(2), 98-106.

Autti-Ramo, I. (2000, June). Twelve-year follow-up of children exposed to alcohol in utero. *Developmental Medicine and Child Neurology, 42*(6), 406-411.

Aronson, M., Hagberg, B., and Gillberg, C. (1997, September). Attention deficits and autistic spectrum problems in children exposed to alcohol during gestation: A follow-up study. *Developmental Medicine and Child Neurology, 39*(9), 583-587.

Oberlander, T. F., Warburton, W., Misri, S., Aghajanian, J., and Hertzman, C. (2006, August). Neonatal outcomes after prenatal exposure to selective serotonin reuptake inhibitor antidepressants and maternal depression using population-based linked health data. *Archives of General Psychiatry, 63*(8), 898-906.

Gray, K. A., Day, N. L., Leech, S., and Richardson, G. A. (2005, May-June). Prenatal marijuana exposure: Effect on child depressive symptoms at ten years of age. *Neurotoxicology and Teratology, 27*(3), 439-448.

Willford, J. A., Richardson, G. A., Leech, S. L., and Day, N. L. (2004). Verbal and visuospatial learning and memory function in children with moderate prenatal alcohol exposure. *Alcoholism: Clinical and Experimental Research, 28*(3), 497-507.

Dunn, C. L., Pirie, P., and Hellerstedt, W. L. (2004). Self exposure to secondhand smoke among prenatal smokers, abstainers, and nonsmokers. *American Journal of Health Promotion, 18*(4), 296-299.

[122] Hutto, D. D. (2008). Folk psychological narratives. *The sociocultural basis of understanding reasons.* Cambridge, MA: MIT Press.

Iacoboni, M. (2008). Mirroring people: The new science of how we connect with others. New York: Farrar, Strauss, and Giroux.

[123] Meltzoff, A. N. (1988). Infant imitation after a 1-week delay: Long-term memory for novel acts and multiple stimuli. *Developmental Psychology, 24*(4), 470-476.

Gergely, G., Bekkering, H., and Király, I. (2002). Rational imitation in preverbal infants. *Nature, 415*(6873), 755.

Andrews, K. (2017). Pluralistic folk psychology in humans and other apes. In J. Kiverstein (Ed.), *The Routledge handbook of philosophy of the social mind*. London: Routledge.

[124] Horner, V. and Whiten, A. (2005). Causal knowledge and imitation/emulation switching in chimpanzees (Pan troglodytes) and children (Homo sapiens). *Animal Cognition, 8*(3), 164-181.

Andrews, K. (2017). Pluralistic folk psychology in humans and other apes. In J. Kiverstein (Ed.), *The Routledge handbook of philosophy of the social mind*. London: Routledge.

[125] Adler, A. (1964b). *Social interest: A challenge to mankind*. New York: Capricorn Books. (Original work published 1929)

[126] Adler, A. (1927). *Understanding human nature*. (W. B. Wolfe, Trans.). Garden City, NY: Garden City Publishing Company, Inc.

[127] Hardin, G. (1968). The tragedy of the commons. *Science, 13, 162*(3859), 1243-1248.

[128] Adler, A. (1996a). The structure of neurosis. *Individual Psychology, 52*(2), 351-362. (Original work published 1935)

[129] Adler, A. (1964b). *Social interest: A challenge to mankind*. New York: Capricorn Books. (Original work published 1929)

Adler, A. (1973). *Superiority and social interest. (3rd. Rev. ed.)*. H. L. Ansbacher, and R. R. Ansbacher, (Eds.). New York: Viking Press.

Bitter, J. R. (1996). On neurosis: An introduction to Adler's concepts and approach. *Individual Psychology: The Journal of Adlerian Theory, Research and Practice, 52*(4), 310-317.

[130] Adler, A. (1964b). *Social interest: A challenge to mankind*. New York: Capricorn Books. (Original work published 1929)

[131] Adler, A. (1956). The Individual Psychology of Alfred Adler: A systematic presentation in selections from his writings. H. Ansbacher and R. Ansbacher (Eds.). New York: Basic Books.

Mosak, H. H. (1991). "I don't have social interest": Social interest as a construct. *Individual Psychology, 47*(3), 309-320.

[132] Gibson, E. J. and Walk, R. D. (1960). The 'visual cliff'. *Scientific American, 202*, 64-71.

[133] Parkinson, B., Phiri, N., and Simons, G. (2012). Bursting with anxiety: Adult social referencing in an interpersonal balloon analogue risk task (BART). *Emotion, 12*(4), 817-826.

Parkinson, B. (2011). Interpersonal emotion transfer: Contagion and social appraisal. *Social and Personality Psychology Compass, 5*(7), 428-439.

Parkinson, B. and Simons, G. (2009). Affecting others: Social appraisal and emotion contagion in everyday decision making. *Personality and Social Psychology Bulletin, 35*(8), 1071-1084.

Parkinson, B. and Simons, G. (2012). Worry spreads: Interpersonal transfer of problem-related anxiety. *Cognition & Emotion, 26*(3), 462-479.

Parkinson, B. (2010). Emotions in interpersonal interaction. In K. R., Scherer, T. Banziger, & E. B. Roesch (Eds.) *Blueprint for affective computing: A sourcebook.* Oxford: Oxford University Press.

[134] Dreikurs, R. and Soltz, V. (1990). *Children: The challenge.* New York: Plume.

[135] Asch, S. E., (1951). Effects of group pressure upon the modification and distortion of judgements. In H. Guetzkow (Ed.) *Groups, Leadership, and Men* (pp. 177-190). Pittsburgh, PA: Carnegie Press.

[136] Bandura, A., Ross, D., and Ross, S. A. (1961). Transmission of aggressions through imitation of aggressive models. *Journal of Abnormal and Social Psychology, 63*(3), 575-582.

[137] Janis, I. (1972). Victims of Groupthink: A psychological study of foreign-policy decisions and fiascoes. Boston: Houghton Mifflin.

[138] Cooley, C. H. (2015). *Human nature and the social order.* Miami, FL: HardPress Publishing. (Original work published 1902)

[139] Milgram, S. (1974). *Obedience to Authority: An Experimental View.* New York: Harper and Row.

[140] Durkheim, E. (1951). *Suicide: A study in sociology.* (J. A. Spaulding & G. Simpson, Trans.). Glencoe, IL: Free Press. (Original work published 1897)

Robertson Blackmore, E., Munce, S., Weller, I., Zagorski, B., Stansfeld, S. A., et al. (2008). Psychosocial and clinical correlates of suicidal acts: Results from a national population survey. *British Journal of Psychiatry, 192*(4), 279-284.

[141] Durkheim, E. (1951). *Suicide: A study in sociology.* (J. A. Spaulding & G. Simpson, Trans.). Glencoe, IL: Free Press. (Original work published 1897)

Luoma, J. B., and Pearson, J. L. (2002). Suicide and marital status in the United States, 1991–1996: is widowhood a risk factor? *American Journal of Public Health, 92*(9), 1518–1522. *Findings support higher suicide rates for widowed persons.

Kposowa, A. J. (2000). Marital status and suicide in the National Longitudinal Mortality Study. *Journal Epidemiology and Community Health, 54*(4), 254–261. *Findings support that divorce has significant correlation with suicide, but only among males.

Popoli, G., Sobelman, S., and Kanarek, N. F. (1989). Suicide in the state of Maryland. Public *Health Reports, 104*(3), 298–301. *Findings show widowed or divorced persons have higher rates of suicide than those single or married.

Smith, J. C., Mercy, J. A., and Conn, J. M. (1988). Marital status and the risk of suicide. American *Journal of Public Health, 78*(1), 78–80. *Findings show that married individuals have lowest suicide rates, with young widowed males having high suicide rates.

[142] Silver, I. M. and Shaw, A. (2018). Pint-sized public relations: The development of reputation management. *Trends in Cognitive Sciences, 22*(4), 277-279.

[143] Aronson, E. (1999). *The social animal*. New York: Worth Publishers.

Insel, B. J. and Gould, M. S. (2008). Impact of modeling on adolescent suicidal behavior. *Psychiatric Clinics of North America, 31*(2), 293-316.

[144] Randell, B. P., Wang, W-L., Herting, J. R., and Eggert, L. L. (2006). Family factors predicting categories of suicide risk. *Journal of Child and Family Studies, 15*(3), 255-270.

[145] Holt-Lunstad, J., Smith, T. B., Baker, M., Harris, T., and Stephenson, D. (2015). Loneliness and social isolation as risk factors for mortality: A meta-analytic review. *Perspectives on psychological science, 10*(2), 227-237.

Holt-Lunstad J., Smith T. B., and Layton J. B. (2010) Social relationships and mortality risk: A meta-analytic review. *PLOS Medicine 7*(7): e1000316. Retrieved from https://doi.org/10.1371/journal.pmed.1000316

[146] Sapolsky, R. M. (1998). Why zebras don't get ulcers: An updated guide to stress, stress-related diseases, and coping. New York: W. H. Freeman and Company.

House, J., Landis, K., and Umberson, D. (1988). Social relationships and health. *Science, 241(4865)*, 540-545.

Berkman, L. F. and Breslow L. (1983). *Health and ways of living: Findings from the Alameda County study*. New York: Oxford University Press.

[147] Pinker, S. (2015). The village effect: How face-to-face contact can make us healthier and happier. Toronto: Vintage Canada.

[148] Ertel, K. A., Glymour, M. M., and Berkman, L. F. (2008). Effects of social integration on preserving memory function in a nationally representative US elderly population. *American Journal of Public Health, 98*(7), 1215-1220.

[149] Holt-Lunstad, J., Smith, T. B., Layton, J. B., and Brayne, C. (2010). Social relationships and mortality risk: A meta-analytic review. *Plos Medicine, 7*(7), e1000316.

[150] Sapolsky, R. M. (1998). Why zebras don't get ulcers: An updated guide to stress, stress-related diseases, and coping. New York: W. H. Freeman and Company.

[151] Sapolsky, R. M. (1998). Why zebras don't get ulcers: An updated guide to stress, stress-related diseases, and coping. New York: W. H. Freeman and Company.

Kiecolt-Glaser, J., Garner, W., Speicher, C., Penn, G., and Glaser, R. (1984). Psychosocial modifiers of immunocompetence in medical students. *Psychosomatic Medicine, 46*(1), 7-14.

Pressman, S. D., Cohen, S., Miller, G. E., Barkin, A., Rabin, B. S., and Treanor, J. J. (2005). Loneliness, social network size, and immune response to influenza vaccination in college freshmen. *Health Psychology, 24*(3), 297–306.

[152] Sapolsky, R. M. (1998). Why zebras don't get ulcers: An updated guide to stress, stress-related diseases, and coping. New York: W. H. Freeman and Company.

Herbert, T., and Cohen, S. (1993). Stress and immunity in humans: A meta-analytic review. *Psychosomatic Medicine, 55*(4), 364-379.

[153] Valtorta, N. K., Kanaan, M., Gilbody, S., Ronzi, S., and Hanratty, B. (2016). Loneliness and social isolation as risk factors for coronary heart disease and stroke: Systematic review and meta-analysis of longitudinal observational studies. *Heart, 102*(13), 1009-1016.

Holt-Lunstad, J. and Smith, T. B. (2016). Loneliness and social isolation as risk factors for CVD: Implications for evidence-based patient care and scientific inquiry. *Heart, 102*(13), 987-989.

Hakulinen C, Pulkki-Råback L, Virtanen M, Jokela, M., Kivimaki, M. et al. (2018). Social isolation and loneliness as risk factors for myocardial infarction, stroke and mortality: UK Biobank cohort study of 479,054 men and women. *Heart.* doi:10.1136/heartjnl-2017-312663

[154] Holt-Lunstad, J., Smith, T. B., Baker, M., Harris, T., and Stephenson, D. (2015). Loneliness and social isolation as risk factors for mortality: A meta-analytic review. *Perspectives on psychological science, 10*(2), 227-237.

Holt-Lunstad J., Smith T. B., and Layton J. B. (2010) Social relationships and mortality risk: A meta-analytic review. *PLOS Medicine 7*(7): e1000316. Retrieved from https://doi.org/10.1371/journal.pmed.1000316

[155] Holmes, T. H. and Rahe, R. H. (1967). The Social Readjustment Rating Scale. *Journal of Psychosomatic Research, 11*(2), 213-218.

[156] Ridley, M. (1999). Genome: The autobiography of a species in 23 chapters. New York: HarperCollins.

[157] Ridley, M. (1999). Genome: The autobiography of a species in 23 chapters. New York: HarperCollins.

[158] Triandis, H. C. (1995). *Individualism & collectivism.* Boulder, CO: Westview Press.

Markus, H. R. and Kitayama, S. (1991). Culture and the self: Implications for cognition, emotion, and motivation. *Psychological review, 98*(2), 224-253.

Markus, H. and Kitayama, S. (1994). The cultural construction of self and emotion: Implications for social behavior. In S. Kitayama, & H. Markus (Eds.), *Emotion and culture* (pp. 89-130). Washington, DC: American Psychological Association.

Cohen, D. and Gunz, A. (2002). As seen by the other . . . : Perspectives on the self in the memories and emotional perceptions of Easterners and Westerners. *Psychological Science, 13*(1), 55-59.

[159] Hughes, M. and Waite, L. (2009, September). Marital biography and health at mid-life. *Journal of Health and Social Behavior, 50*(3), 344-358. Online edition. Retrieved from http://www.asanet.org/galleries/default-file/Sep09JHSBFeature.pdf

[160] Vaillant, G. E. (2003). Aging well: Surprising guideposts to a happier life from the landmark Harvard study of adult development. Boston: Little, Brown and Company.

Vaillant, G. E. (2015). *Triumphs of experience: The men of the Harvard Grant Study.* Cambridge, MA: Belknap Press of Harvard University Press.

[161] Sapolsky, R. M. (1998). Why zebras don't get ulcers: An updated guide to stress, stress-related diseases, and coping. New York: W. H. Freeman and Company.

House, J., Landis, K., and Umberson, D. (1988). Social relationships and health. *Science, 241(4865)*, 540-545.

[162] Sapolsky, R. M. (1998). Why zebras don't get ulcers: An updated guide to stress, stress-related diseases, and coping. New York: W. H. Freeman and Company.

Lepore, S., Allen, K., and Evans, G. (1993). Social support lowers cardiovascular reactivity to an acute stressor. *Psychosomatic Medicine, 55*(6), 518-524.

Edens, J., Larkin, K., and Abel, J. (1992). The effect of social support and physical touch on cardiovascular reactions to mental stress. *Journal of Psychosomatic Research, 36*(4), 371-381.

Gerin, W., Pieper, C., Levy, R., and Pickering, T. (1992). Social support in social interaction: A moderator of cardiovascular reactivity. *Psychosomatic Medicine, 54*(3), 324-336.

Kamarck, T., Manuck, S., and Jennings, J. (1990). Social support reduces cardiovascular reactivity to psychological challenge: A laboratory model. *Psychosomatic Medicine, 52*(1), 42-58.

[163] Taylor S. E. (2011). Affiliation and Stress. In S. Folkman (Ed.). *The Oxford Handbook of Stress, Health and Coping.* (pp. 86-100). New York: Oxford University Press Inc.

Uvnas-Moberg, K. and Petersson, M. (2005). Oxytocin, a mediator of anti-stress, well-being, social interaction, growth and healing. *Z Psychosom Med Psychother. 51*(1), 57-80. Retrieved from http://www.richardhill.com.au/oxytocin.pdf

Fuchs, N. K. (2006). *The health detective's 456 most powerful healing secrets.* Laguna Beach, CA: Basic Health Publications.

Dayton, T. (2007). Emotional sobriety: From relationship trauma to resilience and balance. Deerfield Beach, FL: Health Communications.

[164] Goldstein, P., Weissman-Fogel, I., Dumas, G., and Shamay-Tsoory, S. G. (2018). *Brain-to-brain coupling during handholding is associated with pain reduction.* Proceedings of the National Academy of Sciences, 201703643. Retrieved from https://doi.org/10.1073/ pnas.1703643115

[165] Goldstein, P., Shamay-Tsoory, S. G., Goldstein, P., Goldstein, P., and Weissman-Fogel, I. (2017). The role of touch in regulating inter-partner physiological coupling during empathy for pain. *Scientific Reports, 7*(1), 3252.

[166] Effects of sexual activity on beard growth in man. (1970). *Nature, 226*(5248), 869-870.

Bribiescas, R. G. (2006). *Men: Evolutionary and life history.* Cambridge, MA: Harvard University Press.

[167] Uchino, B. N., Smith, T. W., and Berg, C. A. (2014). Spousal relationship quality and cardiovascular risk: Dyadic perceptions of relationship ambivalence are associated with coronary-artery calcification. *Psychological Science Cambridge,*

25(4), 1037-1042. Retrieved from http://www.ncbi.nlm.nih.gov/pubmed/24501110

[168] Olds, S. W., Marks, L., and Eiger, M. S. (2010). *The complete book of breastfeeding*. New York: Workman Publishing Company.

[169] Powe, C. E., Knott, C. D., and Conklin-Brittain, N. (2010). Infant sex predicts breast milk energy content. *American Journal of Human Biology, 22*(1), 50-54.

Thakkar, S. K., Giuffrida, F., Cruz-Hernandez, C., Castro, C. A., Mukherjee, R., et. al. (2013). Dynamics of human milk nutrient composition of women from Singapore with a special focus on lipids. *American Journal of Human Biology, 25*(6), 770-779.

Fujita, M., Roth, E., Lo, Y.-J., Hurst, C., Vollner, J., et. al. (2012). In poor families, mothers' milk is richer for daughters than sons: A test of Trivers-Willard hypothesis in agropastoral settlements in Northern Kenya. *American Journal of Physical Anthropology, 149*(1), 52-59.

Mandel, D., Lubetzky, R., Dollberg, S., Barak, S., and Mimouni, F. B. (2005). Fat and energy contents of expressed human breast milk in prolonged lactation. *Pediatrics, 116*(3), 432-435.

Hassiotou, F., Hepworth, A. R., Metzger, P., Tat, L. C., Trengove, N., et. al. (2013). Maternal and infant infections stimulate a rapid leukocyte response in breastmilk. *Clinical & Translational Immunology, 2*, e3; doi:10.1038/cti.2013.1

Hassiotou, F. and Geddes, D. T. (2015). Immune cell-mediated protection of the mammary gland and the infant during breastfeeding. *Advances in Nutrition, 6*(3), 267-275.

Sanders, L. (2015). Backwash from nursing babies may trigger infection fighters. *ScienceNews*. Retrieved from https://www.sciencenews.org/blog/growth-curve/backwash-nursing-babies-may-trigger-infection-fighters

[170] Meltzoff, A. N. and Moore, M. K. (1989). Imitation in newborn infants: Exploring the range of gestures imitated and the underlying mechanisms. *Developmental Psychology, 25*(6), 954-962.

Reissland, N. (1988). Neonatal imitation in the first hour of life: Observations in rural Nepal. *Developmental Psychology, 24*(4), 464-69.

Field, T. (1990). *Infancy*. Cambridge, MA: Harvard University Press.

[171] Crozier, W. R. and Jong, P. J. (2012). *The psychological significance of the blush*. Cambridge, UK: Cambridge University Press.

[172] Butler, J. G. (2002). *Television: Critical methods and applications*. Mahwah, NJ: Lawrence Erlbaum Associates.

[173] Hérail, R. J. and Lovatt, E. A. (1996). *Dictionary of modern colloquial French*. (p. 122) New York: Routledge.

[174] Gerst-Emerson, K. and Jayawardhana, J. (2015). Loneliness as a public health issue: The impact of loneliness on health care utilization among older adults. *American Journal of Public Health, 105*(5), 1013-1019.

[175] Feeney, B. C., Van, Vleet. M., Jakubiak, B. K., and Tomlinson, J. M. (2017). Predicting the pursuit and support of challenging life opportunities. *Personality and Social Psychology Bulletin, 43*(8), 1171-1187.

[176] Chopik, W. J. and O'Brien, E. (2017). Happy you, healthy me? Having a happy partner is independently associated with better health in oneself. *Health Psychology, 36*(1), 21-30.

[177] Berscheid E. and Walster E. (1974). Physical attractiveness. In Berkowitz L. (Ed.), *Advances in experimental social psychology* (Vol. 7, pp. 157–215). New York: Academic Press.

Buss D. M. and Barnes M. (1986). Preferences in human mate selection. *Journal of Personality and Social Psychology, 50*(3), 559–570.

Epstein E. and Guttman R. (1984). Mate selection in man: Evidence, theory, and outcome. *Social Biology, 31*(3-4), 243–278.

Spurler J. N. (1963). Assortative mating with respect to physical characteristics. *Eugenics Quarterly, 15*(2), 128–140.

[178] Lee, L., Loewenstein, G., Ariely, D., Hong, J. and Young, J. (2008). If I'm not hot, are you hot or not? Physical-attractiveness evaluations and dating preferences as a function of one's own attractiveness. *Psychological Science, 19(7), 669-677.*

[179] Carroll, J. E., Diez, R. A. V., Fitzpatrick, A. L., and Seeman, T. (2013). Low social support is associated with shorter leukocyte telomere length in late life: Multi-ethnic study of atherosclerosis. *Psychosomatic Medicine, 75*(2), 171-177.

[180] Mampe, B., Friederici, A. D., Christophe, A., and Wermke, K. (2009). Newborns' cry melody is shaped by their native language. *Current Biology, 19*(23), 1994-1997.

[181] Luhrmann, T. M., Padmavati, R., Tharoor, H., and Osei, A. (2015). Differences in voice-hearing experiences of people with psychosis in the USA, India and Ghana: interview-based study. *The British Journal of Psychiatry, 206*(1), 41-44.

[182] Sapolsky, R. M. (1998). Why zebras don't get ulcers: An updated guide to stress, stress-related diseases, and coping. New York: W. H. Freeman and Company.

[183] Freud, S. (1989). *Civilization and its discontents* (p. 33). New York: W. W. Norton and Company. (Original work published 1930)

[184] Adler, A. (1956). The Individual Psychology of Alfred Adler: A systematic presentation in selections from his writings. H. Ansbacher and R. Ansbacher (Eds.). New York: Basic Books.

[185] lunatic. (n.d.). Online Etymology Dictionary. Retrieved from http://www.etymonline.com/ index.php?term=lunatic

[186] Jodelet, D. (1991). Madness and social representations. London: Harvester/Wheatsheaf.

Farr, R.M. (1996). The roots of modern social psychology. Oxford: Blackwell.

Jovchelovitch, S. (2007). Knowledge in context: Representations, community, and culture. London: Routledge.

[187] Knight, J. C. (2010). Human genetic diversity: Functional consequences for health and disease. Oxford: Oxford University Press.

Carey, G. (2003). *Human genetics: For the social sciences.* Thousand Oaks, CA: Sage Publications.

Starr, C., Evers, C. A., and Starr, L. (2015). *Biology: Concepts and applications without physiology.* Stamford, CT: Brooks/Cole.

Hayat, M. A. (2005). Handbook of immunohistochemistry and in situ hybridization of human carcinomas: Volume 3. Boston: Elsevier Academic Press.

[188] Mosak, H. H. (1972). Life style assessment: A demonstration focused upon family constellation. *Journal of Individual Psychology, 28*(2), 232-247.

Shulman, B. H. and Mosak, H. H. (1988). *A manual of life style assessment.* Muncie, IN: Accelerated development.

Withrow, R and Schwiebert, V. L. (2005). Twin loss: Implications for counselors working with surviving twins. *Journal of Counseling and Development, 83*(1), 21-28.

[189] Deci, E. L. and Ryan, R. M. (1985). *Intrinsic motivation and self-determination in human behavior.* New York: Plenum Press.

[190] Deci, E. L. (1980). *The psychology of self-determination.* Lexington, MA: Lexington Books.

[191] Gookin, D. (2008). *Troubleshooting your PC for dummies (3rd ed.).* Hoboken, NJ: John Wiley & Sons.

[192] Moore, D. S. (2003). The dependent gene: The fallacy of nature vs. nurture. New York: Henry Holt.

[193] Ryback, R. (2016, January 4). Psychiatrist vs. psychologist: What to know before seeking one of these professionals. *Psychology today.* Retrieved from https://www.psychologytoday.com /blog/the-truisms-wellness/201601/psychiatrist-vs-psychologist

American Psychiatric Association (2018). What is psychiatry? *American Psychiatric Association.* Retrieved from https://www.psychiatry.org/patients-families/what-is-psychiatry

Cassata, C. (n.d.). What is a psychiatrist? *Everyday health.* Retrieved from https://www.everydayhealth.com/psychiatrist/guide/

[194] Rubin, E. (2010, May 17). *Is the training of a psychiatrist more like that of a neurologist or a psychologist? Psychology today.* Retrieved from https://www.psychologytoday.com/ blog/demystifying-psychiatry/201005/is-the-training-psychiatrist-more-neurologist-or-psychologist

[195] Pomerantz, A. M. (2017). *Clinical psychology: Science, practice, and culture.* Los Angeles, CA: Sage Publications.

Rubin, E. (2010, May 17). *Is the training of a psychiatrist more like that of a neurologist or a psychologist? Psychology today.* Retrieved from https://www.psychologytoday.com/ blog/demystifying-psychiatry/201005/is-the-training-psychiatrist-more-neurologist-or-psychologist

American Psychiatric Association (2018). What is psychiatry? *American Psychiatric Association*. Retrieved from https://www.psychiatry.org/patients-families/what-is-psychiatry

[196] Pomerantz, A. M. (2017). *Clinical psychology: Science, practice, and culture.* Los Angeles, CA: Sage Publications.

Rubin, E. (2010, May 17). *Is the training of a psychiatrist more like that of a neurologist or a psychologist? Psychology today.* Retrieved from https://www.psychologytoday.com/ blog/demystifying-psychiatry/201005/is-the-training-psychiatrist-more-neurologist-or-psychologist

[197] Bitter, J. R. (1996). On neurosis: An introduction to Adler's concepts and approach. *Individual Psychology: The Journal of Adlerian Theory, Research and Practice, 52*(4), 310-317.

Sperry, L. (1996). Adler's conception of neurosis in the contemporary treatment of personality disorders. *Individual Psychology: The Journal of Adlerian Theory, Research and Practice, 52*(4), 372-377.

Mozdzierz, G. J. (1996). Adler's "What is neurosis?": Clinical and predictive revelations from the past. *Individual Psychology: The Journal of Adlerian Theory, Research and Practice, 52*(4), 342-350.

[198] Adler, A. (1956). The Individual Psychology of Alfred Adler: A systematic presentation in selections from his writings. H. Ansbacher and R. Ansbacher (Eds.). New York: Basic Books.

[199] Mosak, H. H. (1958). Early recollections as a projective technique. *Journal of Projective Techniques, 22*(3), 302–311.

[200] Mosak, H. H. and Maniacci, M. P. (1999). A primer of Adlerian psychology: The analytic-behavioral-cognitive psychology of Alfred Adler. Philadelphia, PA: Brunner/Mazel.

[201] Boldero, J., Moretti, M., Bell, R., and Francis, J. (2005). Self-discrepancies and negative affect: A primer on when to look for specificity, and how to find it. *Australian Journal of Psychology, 57*(3), 139-147.

Coon, D., Mitterer, J. O., Talbot, S., and Vanchella, C. M. (2010). *Introduction to psychology: Gateways to mind and behavior.* Belmont, CA: Wadsworth Cengage Learning.

[202] Adler, A. (1956). The Individual Psychology of Alfred Adler: A systematic presentation in selections from his writings. H. Ansbacher and R. Ansbacher (Eds.). New York: Basic Books.

[203] Adler, A. (1958). *What life should mean to you.* A. Porter (Ed.). New York: Capricorn Books. (Original work published 1931)

[204] Spiegler, M. D. and Guevremont, D. C. (2010). *Contemporary behavior therapy.* Belmont, CA: Wadsworth, Cengage Learning.

Martin, G. and Pear, J. J. (2014). *Behavior modification: What it is and how to do it*, Tenth Edition. Boston: Pearson Education.

[205] Krausz, E. O. (1973). Neurotic versus normal reaction categories. In H. H. Mosak (Ed.). *Alfred Adler: His influence on psychology today* (pp. 53-57). Park Ridge, NJ: Noyes Press.

[206] Sartre, J.–P. (1989). *No exit and three other plays* (p. 45). New York: Vintage International. (Original work published 1946)

[207] Darwin, C. (1859). On the origin of species by means of natural selection, or the preservation of favoured races in the struggle for life. London: John Murray.

Dawkins, R. (1989). *The selfish gene.* New York: Oxford University Press.

Kemp, T. S. (2005). *The origin and evolution of mammals.* New York: Oxford University Press.

Niklas, K. J. (1997). The *evolutionary biology of plants.* Chicago: University of Chicago Press.

Pearson, L. C. (1995). *The diversity and evolution of plants.* Boca Raton, FL: CRC Press Inc.

Williams, G. C. (1966). *Adaptation and natural selection.* Princeton: Princeton University Press. As cited in Dawkins, R. (1989). *The selfish gene.* New York: Oxford University Press.

[208] Dawkins, R. (1989). *The selfish gene.* New York: Oxford University Press.

Dawkins, R. (1996). *The blind watchmaker.* New York: W. W. Norton.

[209] Barrett, D. B., Kurian, G. T., and Johnson, T. M. (Eds.). (2001). World Christian Encyclopedia: A comparative survey of churches and religions in the modern world: Volumes 1-2. Oxford: Oxford University Press.

Ostling, R. N. (2001) *Researcher tabulates world's believers.* (Originally published in the Salt Lake Tribune, from http://www.adherents.com/misc/WCE.html).

[210] Adler, A. (1958). *What life should mean to you.* A. Porter (Ed.). New York: Capricorn Books. (Original work published 1931)

[211] Adler, A. (1958). *What life should mean to you.* A. Porter (Ed.). New York: Capricorn Books. (Original work published 1931)

[212] Adler, A. (1929). *Practice and theory of individual psychology* (Rev. ed.). London: Lund Humphries.

Adler, A. (1956). The Individual Psychology of Alfred Adler: A systematic presentation in selections from his writings. H. Ansbacher & R. Ansbacher (Eds.), New York: Basic Books.

Mosak, H. H. and Maniacci, M. P. (1999). A primer of Adlerian psychology: The analytic-behavioral-cognitive psychology of Alfred Adler. Philadelphia, PA: Brunner/Mazel.

Rasmussen, P. R. (2015). *The quest to feel good.* New York: Routledge.

Dinkmeyer, D. C., Pew, W. L., and Dinkmeyer, D. C. (1979). *Adlerian counseling and psychotherapy*. Monterey, CA: Brooks/Cole Pub. Co.

Chew, A. L. (1998). A primer on Adlerian psychology: Behavior management techniques for young children. Atlanta, GA: Humanics Trade.

Hall, C. S. and Lindzey, G. (1970). *Theories of personality* (2nd ed.) New York: John Wiley & Sons.

Beams, T. B. (1992). *A student's glossary of Adlerian terminology* (2nd Ed.) Ladysmith, BC: Photon Communications.

[213] Milliren, A., Clemmer, F., Wingett, W., and Testerment, T. (2005). The movement from "felt minus" to "perceived plus": Understanding Adler's concept of inferiority (pp. 351-363). In S. Slavik & J. Carlson, (Eds.), *Readings in the theory of Individual Psychology*. New York: Routledge.

Mosak, H. H. and Maniacci, M. P. (1999). A primer of Adlerian psychology: The analytic-behavioral-cognitive psychology of Alfred Adler. Philadelphia, PA: Brunner/Mazel.

[214] Adler, A. (1996a). The structure of neurosis. *Individual Psychology, 52*(2), 351-362. (Original work published 1935)

Milliren, A., Clemmer, F., Wingett, W., and Testerment, T. (2005). The movement from "felt minus" to "perceived plus": Understanding Adler's concept of inferiority (pp. 351-363). In S. Slavik & J. Carlson, (Eds.), *Readings in the theory of Individual Psychology*. New York: Routledge.

Mosak, H. H. and Maniacci, M. P. (1999). A primer of Adlerian psychology: The analytic-behavioral-cognitive psychology of Alfred Adler. Philadelphia, PA: Brunner/Mazel.

[215] Adler, A. (1939). Sur la "protestation virile." [On the "masculine protest"]. Courage, 2(1). 8-10. As cited in Mosak, H. H. and Maniacci, M. P. (1999). *A primer of Adlerian psychology: The analytic-behavioral-cognitive psychology of Alfred Adler*. Philadelphia, PA: Brunner/Mazel.

Mosak, H. H. and Schneider, S. (1989). Masculine protest, penis envy, women's liberation and sexual equality. *Journal of Individual Psychology, 33*(2), 193-202.

Mosak, H. H. and Maniacci, M. P. (1999). A primer of Adlerian psychology: The analytic-behavioral-cognitive psychology of Alfred Adler. Philadelphia, PA: Brunner/Mazel.

[216] Chase-Dunn, C. and Lerro, B. (2014). *Social change: Globalization from the stone age to the present*. Boulder, CO: Paradigm.

Chase-Dunn, C. (2016). The socio-cultural evolution of world-systems. In J. H., Turner, R. Machalek, & A. Maryanski (Eds.). *Handbook on evolution and society: Toward an evolutionary social science*. New York: Routledge.

[217] Adler, A. (1958). *What life should mean to you*. A. Porter (Ed.). New York: Capricorn Books. (Original work published 1931)

[218] Lundin, R. W. (1989). *Alfred Adler's basic concepts and implications.* Muncie, IN: Accelerated Development Inc.

[219] Adler, A. (1956). The Individual Psychology of Alfred Adler: A systematic presentation in selections from his writings. H. Ansbacher and R. Ansbacher (Eds.). New York: Basic Books.

Smuts, J. C. (1973). *Holism and evolution.* Oxford: Greenwood Press.

[220] Kahneman, D. (2015). *Thinking, fast and slow.* New York: Farrar, Straus and Giroux.

Zajonc, R. B. (2016). Mere exposure: A gateway to the subliminal. *Current Directions in Psychological Science, 10*(6), 224-228.

Veritasium. (2016, July 21). *The illusion of truth.* Retrieved from https://www.youtube.com/ watch?v=cebFWOlx848

[221] Adler, A. (1929). *Practice and theory of individual psychology* (Rev. ed.). London: Lund Humphries.

Adler, A. (1931). *What life should mean to you.* Boston: Little, Brown and Company.

Adler, A. (1956). The Individual Psychology of Alfred Adler: A systematic presentation in selections from his writings. H. Ansbacher and R. Ansbacher (Eds.). New York: Basic Books.

Shulman, B. H. (1977a). Encouraging the pessimist: A confronting technique. *The Individual Psychologist, 14*(1), 7-9.

Shulman, B. H. and Dreikurs, S. G. (1978). The contributions of Rudolf Dreikurs to the theory and practice of Individual Psychology. *Journal of Individual Psychology, 34*(2), 153-169.

Griffith, J. and Powers, B. (1984). An Adlerian lexicon: Fifty-nine terms associated with the Individual Psychology of Alfred Adler. Chicago: The Americas Institute of Adlerian Studies.

Rasmussen, P. R. (2015). *The quest to feel good.* New York: Routledge.

Beams, T. B. (1992). *A student's glossary of Adlerian terminology* (2nd Ed.) Ladysmith, BC: Photon Communications.

[222] Adler, A. (1931). *What life should mean to you.* Boston: Little, Brown.

Beams, T. B. (1992). *A student's glossary of Adlerian terminology* (2nd Ed.) Ladysmith, BC: Photon Communications.

[223] Shulman, B. H. (1977a). Encouraging the pessimist: A confronting technique. *The Individual Psychologist, 14*(1), 7-9.

[224] Amazon. com, Inc. (n.d.). Amazon.com: Help > Amazon.com Site Features > Your Content > Recommendations. Retrieved from http://www.amazon.com/gp/help/customer/ display.html?nodeId=13316081#how

Apple, Inc. (2008, September 9). *Apple special event, September 2008.* Retrieved from http://www.apple.com/quicktime/qtv/letsrock/

Linden, G., Smith, B., and York, J. (2003, Jan/Feb). Amazon.com recommendations: Item-to-item collaborative filtering. *IEEE Internet computing, 7*(1), 76-80.

[225] Watson, J. and Crick, F. (1953). Molecular structure of nucleic acids: A structure for deoxyribose nucleic acid. *Nature, 171* (4356), 737–738. Retrieved from http://profiles.nlm.nih.gov/SC/B/B/Y/W/_/scbbyw.pdf

[226] Mosak, H. H. and Maniacci, M. P. (2006). Of cookie jars and candy bars: Dysthymia in the light of Individual Psychology. *The Journal of Individual Psychology, 62*(4), 357-365.

Mosak, H. H. (1977). The interrelatedness of the neuroses through central themes. In H. H. Mosak (Ed.). *On purpose*. Chicago: Adler School of Professional Psychology. (Originally published in Journal of Individual Psychology, 1968, 24, 67-70.)

[227] Mosak, H. H. and Maniacci, M. P. (2006). Of cookie jars and candy bars: Dysthymia in the light of Individual Psychology. *The Journal of Individual Psychology, 62*(4), 357-365.

Mosak, H. H. (1977). The interrelatedness of the neuroses through central themes. In H. H. Mosak (Ed.). *On purpose*. Chicago: Adler School of Professional Psychology. (Originally published in Journal of Individual Psychology, 1968, 24, 67-70.)

[228] Mosak, H. H. and Maniacci, M. P. (2006). Of cookie jars and candy bars: Dysthymia in the light of Individual Psychology. *The Journal of Individual Psychology, 62*(4), 357-365.

Mosak, H. H. (1977). The interrelatedness of the neuroses through central themes. In H. H. Mosak (Ed.). *On purpose*. Chicago: Adler School of Professional Psychology. (Originally published in Journal of Individual Psychology, 1968, 24, 67-70.)

[229] Mosak, H. H. (1977). The interrelatedness of the neuroses through central themes. In H. H. Mosak (Ed.). *On purpose*. Chicago: Adler School of Professional Psychology. (Originally published in Journal of Individual Psychology, 1968, 24, 67-70.)

Mosak, H. H. and Maniacci, M. P. (2006). Of cookie jars and candy bars: Dysthymia in the light of Individual Psychology. *The Journal of Individual Psychology, 62*(4), 357-365.

[230] Mosak, H. H. (1977). The interrelatedness of the neuroses through central themes. In H. H. Mosak (Ed.). *On purpose*. Chicago: Adler School of Professional Psychology. (Originally published in Journal of Individual Psychology, 1968, 24, 67-70.)

Mosak, H. H. and Maniacci, M. P. (2006). Of cookie jars and candy bars: Dysthymia in the light of Individual Psychology. *The Journal of Individual Psychology, 62*(4), 357-365.

[231] Mosak, H. H. and Maniacci, M. P. (2006). Of cookie jars and candy bars: Dysthymia in the light of Individual Psychology. *The Journal of Individual Psychology, 62*(4), 357-365.

[232] Mosak, H. H. and Maniacci, M. P. (2006). Of cookie jars and candy bars: Dysthymia in the light of Individual Psychology. *The Journal of Individual Psychology, 62*(4), 357-365.

[233] Adler, A. (1935). The fundamental views of Individual Psychology. *International Journal of Individual Psychology, 1*(1), pp. 5-8.

Mosak, H. H. (1977). The controller: A social interpretation of the anal character. In H. H. Mosak (Ed.). *On purpose* (pp. 216-227). Chicago: Alfred Adler Institute. (Originally published in Alfred Adler: His influence on psychology today, 1973. Park Ridge, NJ: Noyes Press.)

[234] Mosak, H. H. (1977). The controller: A social interpretation of the anal character. In H. H. Mosak (Ed.). *On purpose* (pp. 216-227). Chicago: Alfred Adler Institute. (Originally published in Alfred Adler: His influence on psychology today, 1973. Park Ridge, NJ: Noyes Press.)

[235] Mosak, H. H. (1977). The controller: A social interpretation of the anal character. In H. H. Mosak (Ed.). *On purpose* (pp. 216-227). Chicago: Alfred Adler Institute. (Originally published in Alfred Adler: His influence on psychology today, 1973. Park Ridge, NJ: Noyes Press.)

[236] Mosak, H. H. (1977). The controller: A social interpretation of the anal character. In H. H. Mosak (Ed.). *On purpose* (pp. 216-227). Chicago: Alfred Adler Institute. (Originally published in Alfred Adler: His influence on psychology today, 1973. Park Ridge, NJ: Noyes Press.)

[237] Mohney, G. (2013, May 25). Motorists can't face fears, get a lift across bridge. *ABC News*. Retrieved from http://abcnews.go.com/Health/terrified-motorists-lift-bridge/story?id= 19250164

[238] Mosak, H. H. (1977). The controller: A social interpretation of the anal character. In H. H. Mosak (Ed.) *On purpose* (pp. 216-227). Chicago: Alfred Adler Institute. (Originally published in Alfred Adler: His influence on psychology today, 1973. Park Ridge, NJ: Noyes Press.)

[239] Mosak, H. H. and Maniacci, M. P. (2006). Of cookie jars and candy bars: Dysthymia in the light of Individual Psychology. *The Journal of Individual Psychology, 62*(4), 357-365.

[240] Mosak, H. H. and LeFevre, C. (1976). The resolution of "intrapersonal" conflict. *Journal of Individual Psychology, 32*(1), 19-26.

[241] Hewitt, P. L., Mittelstaedt, W. M., and Flett, G. L. (1990). Self-oriented perfectionism and generalized performance importance in depression. *Individual Psychology: The Journal of Adlerian Theory, Research and Practice, 46*(1), 67-73.

[242] Dreikurs, R. (1953). *Fundamentals of Adlerian psychology*. Chicago: Alfred Adler Institute. (Original work published 1935)

Orgler, H. (1963). *Alfred Adler: The man and his work*. New York: Capricorn Books. (Original work published 1939)

Mosak, H. H. and Maniacci, M. P. (1999). A primer of Adlerian psychology: The analytic-behavioral-cognitive psychology of Alfred Adler. Philadelphia, PA: Brunner/Mazel.

[243] Adler, Alexandra. (1938). Guiding human misfits: A practical application of Individual Psychology. New York: Macmillan.

[244] Mosak, H. H. and Maniacci, M. P. (2006). Of cookie jars and candy bars: Dysthymia in the light of Individual Psychology. *The Journal of Individual Psychology, 62*(4), 357-365.

[245] Adler, A. (1926). The neurotic constitution: Outlines of a comparative individualistic psychology and psychotherapy. (Bernard Glueck & John E. Lind, Trans.) New York: Dodd, Mead and Company.

[246] Way, L. M. (1950). *Adler's place in psychology*. London: George Allen and Unwin LTD.

[247] Adler, A. (1958). *What life should mean to you*. A. Porter (Ed.). New York: Capricorn Books. (Original work published 1931)

Printed in Great Britain
by Amazon